## HANDLING CLIMATE DISPLACEMENT

How do we begin to handle the greatest crisis affecting humanity today? Climate change is already causing droughts, flooding and famine, forcing people to leave their livelihoods and communities. In the years to come, millions will find their local areas uninhabitable, as mass displacement of people reaches disastrous levels. *Handling Climate Displacement* explains how climate change has become recognized as a human rights concern, and how human rights are key to managing the crisis. Local authorities and populations increasingly call for guidance in the absence of an internationally recognized framework. Dr. Dr. Khaled Hassine LL.M. uses the Peninsula Principles on Climate Displacement within States, drafted in 2013 by a committee of experts and practitioners, to offer concrete solutions to the impending emergency. Enriched by his experience working with the victims of climate displacement, this book offers an effective framework to deal with the challenges presented by mass displacement while protecting human rights.

Khaled Hassine specializes in international and human rights law and has worked extensively on housing, land and property (HLP) rights and redress mechanisms, as well as on land solutions for climate displaced persons. He has advised UN peace operations, research institutes and NGOs, and he is a member of the HLP Group of the UN Global Protection Cluster and the UNHCR Livelihoods Advisory Board. He was honored with an Academia award for his outstanding contribution to research on HLP rights and mass claims procedures in post-conflict situations. He currently works at the UN Secretariat with the UN Human Rights Office.

# Handling Climate Displacement

KHALED HASSINE

CAMBRIDGE
UNIVERSITY PRESS

**CAMBRIDGE**
**UNIVERSITY PRESS**

University Printing House, Cambridge CB2 8BS, United Kingdom

One Liberty Plaza, 20th Floor, New York, NY 10006, USA

477 Williamstown Road, Port Melbourne, VIC 3207, Australia

314–321, 3rd Floor, Plot 3, Splendor Forum, Jasola District Centre, New Delhi – 110025, India

79 Anson Road, #06–04/06, Singapore 079906

Cambridge University Press is part of the University of Cambridge.

It furthers the University's mission by disseminating knowledge in the pursuit of education, learning, and research at the highest international levels of excellence.

www.cambridge.org
Information on this title: www.cambridge.org/9781108486484
DOI: 10.1017/9781316999264



First published 2019

Printed and bound in Great Britain by Clays Ltd, Elcograf S.p.A.

*A catalogue record for this publication is available from the British Library.*

*Library of Congress Cataloging-in-Publication Data*

NAMES: Hassine, Khaled, author.
TITLE: Handling climate displacement / Khaled Hassine, United Nations, Geneva.
DESCRIPTION: Cambridge, United Kingdom ; New York, NY : Cambridge University Press, [2019] | Includes index.
IDENTIFIERS: LCCN 2019014846 | ISBN 9781108486484
SUBJECTS: LCSH: Climatic changes – Law and legislation. | Environmental refugees – Legal status, laws, etc. | Forced migration. | United Nations Framework Convention on Climate Change (1992 May 9). Protocols, etc. (2015 December 12) | Environmental law, International. | Greenhouse gas mitigation – Law and legislation. | Human rights.
CLASSIFICATION: LCC K3585.5 .H37 2019 | DDC 341.4/86–dc23
LC record available at https://lccn.loc.gov/2019014846

ISBN 978-1-108-48648-4 Hardback

*Nihil de nobis, sine nobis*

We lost our occupation,
which means the confidence
that we are of some use in this world.
We lost our language,
which means the naturalness of reactions,
the simplicity of gestures,
the unaffected expression of feelings. [. . .]
Our identity is changed so frequently
that nobody can find out who we actually are.
[. . .], and that means the rupture of our private lives.

Hannah Arendt, "We Refugees"

*Foranid*

# Contents

# Foreword

Climate change is one of the greatest challenges facing the world today. The individuals and communities displaced from their homes and lands as a result of climate change are the human face of this tragedy. Bangladesh, for instance, suffers from regular natural hazards, including floods, tropical cyclones, storm surges and droughts as one of the most vulnerable countries in the world. These hazards are already leading to the loss and destruction of housing, land and property, the loss of livelihoods and widespread displacement across the country. Every year, tens of thousands of people across the country are already losing their homes to rising sea levels, violent tropical cyclones, devastating floods and other impacts of climate change. Climate change in Bangladesh has started what may become the largest mass displacement and migration in human history. Of Bangladesh's sixty-four districts, twenty-six coastal and mainland districts are already producing climate displacement. The country will have to face the challenge of mass displacement, both internally and externally, due to climate change as the country is not yet adequately prepared in providing permanent rights-based solutions through relocation to such a large number of climate displaced people.

Young Power in Social Action (YPSA) and Displacement Solutions have been jointly implementing the Bangladesh Housing, Land and Property Rights Initiative to raise the profile of the climate displacement issue in the country since 2011. Through the Initiative, YPSA and Displacement Solutions have developed an in-depth understanding of climate displacement through consultations with climate displaced persons and through extensive travel across Bangladesh, including fact-finding and research visits to virtually all the climate affected districts throughout the country. The Initiative has achieved a number of considerable successes, including comprehensive training sessions to over 5,000 civil society representatives and government officials on the housing, land and property rights of climate displaced persons, as well as research and publications, and advocacy and lobbying with local, regional and national government officials on the need to respond to climate displacement in a rights-based framework.

To solve these problems in line with a rights-based approach, YPSA, as a civil society organization representative of Bangladesh, actively participated in developing the Peninsula Principles on Climate Displacement within States (2013), outlining the rights of people and communities who lose their homes, land and livelihoods due to the effects of climate change. These Peninsula Principles provide a comprehensive normative framework, based on principles of international law, human rights obligations and good practice, within which the rights of climate displaced persons can be addressed. They address climate displacement within a State and not cross-border climate displacement. This framework sets out protection and assistance principles, consistent with the United Nations Guiding Principles on Internal Displacement, to be applied to climate displaced persons.

As a result of successful advocacy at different levels by YPSA and other like-minded organizations, the Bangladeshi government has already incorporated the displacement issue in many relevant documents and drafted a strategy titled National Strategy on the Management of Disaster and Climate Induced Internal Displacement under the Comprehensive Disaster Management Programme (CDMP II) of the Ministry of Disaster Management and Relief. The strategy broadly addresses the many rights-based solutions in line with the Peninsula Principles.

This study conducted by my friend Khaled Hassine comprehensively analyzed existing international principles, including the Peninsula Principles, to ascertain rights-based solutions to climate displacement in the absence of an internationally agreed framework. I am confident that this book will greatly contribute to ensuring rights-based solutions to climate displacement and that the world will increasingly understand the true dimensions of climate displacement, where it occurs and the immense human suffering inevitably associated with this growing phenomenon.

**Md. Arifur Rahman**
***Chief Executive***
***Young Power in Social Action***
***Bangladesh***

# Preface

Though still far too underappreciated and taken for granted by many, the international legal order – in particular the human rights elements thereof – is surely one of the most profound advancements of the post-World War II period. The system may remain overly easy to evade for violators and difficult to enforce by victims, allowing grotesque human rights criminals to live carefree lives of almost assured impunity, but despite its flaws, human rights law has fundamentally altered the world for the better in nothing less than extraordinary ways. As this legal domain has evolved over the decades, its corpus of law has expanded into areas that simply were not envisaged at the time our global human rights era commenced in 1945. Two of these domains – housing, land and property rights, and rights relating to the environment – have developed more than most, due in part to growing assaults on these rights just about everywhere. And it is the merger of these two areas in the manifestation of climate displacement that is addressed in this remarkable book by the dedicated international lawyer and human rights advocate Dr. Khaled Hassine.

The consequences of human-generated climate change on this solitary planet we all share commenced with the onset of the Industrial Revolution more than 200 years ago. Since that time, while our economies have grown, technology has progressed and many human comforts have increased, the lasting impacts of the economic choices we as humanity have chosen for the past two centuries may have improved many things, but in the process have instigated forces that are now rapidly destroying the very basis of life on the place we all share: Earth. Without even considering the more than hundred species that become extinct *every single day*, the millions of acres of primary forests decimated each year, the tens of millions of people who die annually from pollution, the plasticization of our oceans and the countless other so-called externalities of the global marketplace, the effects of burning nonrenewable fossil fuels upon the climate of our world has been staggering. As $CO_2$ levels in the atmosphere continue to climb, and as average global temperature levels head skyward, the sea level has begun to rise and will continue to rise even if every country were to comply fully with the terms of the 2015 Paris Climate Agreement.

And these effects on humans are horrendous. I have seen it with my own eyes so many times, in so many places, that each passing day without more concerted action to support these victims sends shivers down my spine. We live in a world where the institutions, states and powers still fail to see and act adequately in response to the human misery directly caused by climate change. Yes, the United Nations and its agencies are slowly grasping the scale of this crisis, and many governments, particularly those most heavily affected, are starting to do something, but we are still nowhere near having the laws and procedures in place that we need to fully protect the rights of climate change's victims, in particular those displaced because of it.

The norms and principles are there, for instance in the form of the Peninsula Principles on Climate Displacement within States (2013), which I had the honor to work on developing together with the author of this book, but standards such as these are still yet to sufficiently develop into concrete actions everywhere they are needed. The Peninsula Principles have inspired national climate policies in Bangladesh, Fiji, Panama, the Solomon Islands and elsewhere, have been addressed by the United Nations and are the subject of numerous books and reports, but in real terms their core promise to climate displaced persons everywhere that no one should end up homeless or landless because of enduring climate displacement remains unfulfilled.

And yet the basic parameters of what needs to be done in all states affected, if human rights are to guide this process, are crystal clear, not always easy or cheap, but invariably doable *everywhere*. For if we believe that a future of more and worse slums is one to look forward to, then we do nothing. We effectively say to the climate displaced: "You are free to move. You are on your own. Good luck finding a new home." Is this really the best we can do to address one of the world's largest and rapidly growing problems? Are we really ready to accept that some of the world's already most vulnerable people are to again be left to their own devices and simply accept that people literally losing everything, including their land, is the price we pay for economic growth and skewed, inequitable prosperity?

I know that for people like Khaled Hassine, myself and millions upon millions of others, we are not ready to accept that view. We believe that law, and human rights laws in particular, must form the basis for preventing and resolving climate displacement, and that only through rights-based, planned, people-driven and adequately resourced programs will climate displacement be sufficiently addressed. We have the resources needed to do this. We have the land needed to do this. We have the field-tested experience to do this. Now we need the political will to do so.

This book provides the depth, the arguments and the analysis we need to build this political will into a powerful force grounded in law that will assist in constructing a global coalition of states, organizations and civil society, that yields new laws and fully financed policies that protect now and forever the hundreds of millions of people who stand to lose their homes and lands because of the economic choices we have collectively made.

**Scott Leckie**
***Founder and Director***
***Displacement Solutions***

# Acknowledgments

I am immensely indebted to Professor Dr. Otmar Höll, who, without any hesitation, assumed the supervision of my research project on handling climate displacement. Due to his long-standing experience working at the intersection of research and politics he was best placed to assess this applied research project. I would like to thank him for his openness and guidance. My sincere gratitude also goes to Professor Dr. Helmut Kramer as his support was indispensable in finalizing this project. Finally, I am tremendously grateful to Professor DDr. Ingfrid Schütz-Müller, who accompanied and encouraged this project for many years. He fortified my belief that "*academic research and practical experience are . . . two sides of the same coin.*"[1]

My friend Scott Leckie brought me on board as part of the team of experts who endorsed the Peninsula Principles on Climate Displacement within States. Like him, I had first specialized in housing, land and property rights and later developed a keen interest in addressing the gap of climate displacement. The fact that the Peninsula Principles emerged through a push from grassroots organizations enticed me to analyze them in more depth from an academic angle, trying to join theory and practice. While, through my involvement, I certainly have views about the Peninsula Principles, I approached this research open-mindedly without intransigent preconceived ideas about its outcome. This research was about letting the scientific argument take place, and collecting and analyzing evidence that would confirm or refute my own judgment.

My gratitude goes to all the experts – and friends indeed – who unconditionally shared with me their firsthand experience in dealing with climate displaced communities and their invaluable insight.

Family was indispensable in this endeavor; *Anidleaninafaf* my constant source of inspiration.

[1] I. Schütz-Müller, Excursions to UN Headquarters and Specialized Agencies: Report on Experiences, 2009, p. 106.

# Author's Note

The views expressed in this publication are those of the author and do not necessarily reflect the views of the United Nations Secretariat or any other organization the author was previously affiliated with (ST/IC/2006/30, paras. 7 and 8).

# Requiescat in Pace

## *Prof. DDr. Ingfrid Schütz-Müller*

*Ein Leben für die Internationalität*
*A life for internationality*

IN MEMORIAM

With the passing of Ingfrid Schütz-Müller, we lost an outstanding, admired scholar and inspirational figure who knew that human beings acting together could make the world a better place. He dedicated his life to this belief. He was convinced that the academia and actors have to close ranks, come together and exchange. He provided the – still – much needed space and supported research at the frontiers of knowledge. He encouraged applied research as well as the development of innovative approaches and concrete solution pathways. He embodied the ideas he worked for, continued to motivate and inspire generations of inquiring minds and, thus, created an esprit for international dialogue, thought and creativity in Vienna.

# Abbreviations

| | |
|---|---|
| ARA-JPD | Applied Research Association on Justice, Peace and Development |
| CAF | Cancún Adaptation Framework |
| CESCR | Committee on Economic, Social and Cultural Rights |
| cf. | confer |
| COP | Conference of the Parties |
| ECOSOC | United Nations Economic and Social Council |
| et al. | et alia |
| et subseq. | et subsequent |
| e.g. | exempli gratia |
| i.e. | id est |
| IASC | Inter-Agency Standing Committee |
| ibid. | ibidem |
| ICCPR | International Covcnant on Civil and Political Rights |
| ICESCR | International Covenant on Economic, Social and Cultural Rights |
| IDPs | internally displaced persons |
| IOM | International Organization for Migration |
| IPCC | Intergovernmental Panel on Climate Change |
| n.d. | no date |
| NGO | nongovernmental organization |
| OCHA | United Nations Office for the Coordination of Humanitarian Affairs |
| OECD | Organisation for Economic Co-operation and Development |
| OHCHR | Office of the United Nations High Commissioner for Human Rights |
| SIDS | Small Island Developing States |
| UDHR | Universal Declaration of Human Rights |
| UNDP | United Nations Development Programme |
| UNEP | United Nations Environment Programme |
| UNFCCC | United Nations Framework Convention on Climate Change |
| UNHCR | United Nations High Commissioner for Refugees |

| | |
|---|---|
| UNISDR | United Nations International Strategy for Disaster Risk Reduction |
| viz. | videlicet |
| YPSA | Young Power in Social Action |
| WIM | Warsaw International Mechanism |

# Prologue

Working with the United Nations in Kosovo in 1999, I had witnessed the despair of displaced populations in the aftermath of conflict – their little prospect for return and for recovery of their housing, land and property. I soon became involved in the establishment of the first international mechanism – the Housing and Property Directorate/Claims Commission (HPD/CC) – that would adjudicate such individual housing, land and property claims. Scott Leckie, then Director of the Centre on Housing Rights and Evictions (COHRE), had designed the blueprint for this unique claims settlement mechanism,[1] and the farsighted UN leadership – notably Dan Lewis – made it possible to set this important precedent for the handling of displacement-related claims. From there on, it would become a recurring and dominant topos of my research and professional career, responding to requests for expert guidance in places such as Yemen, Georgia and the Sudan.

Scott's and my collaboration on the Commentary of the United Nations Principles on Housing and Property Restitution for Refugees and Displaced Persons, colloquially known as the Pinheiro Principles,[2] gave the impetus for this research project on handling climate displacement. At that time, Scott had already created Displacement Solutions, shifting his focus to what he considered to be the most significant challenge of our times: climate displacement. His organization offered the ideal institutional base for my research project. In view of the lack of international frameworks, I initially started to analyze how far the Pinheiro Principles could help bridge the protection gap faced by climate displaced persons. When the urge by local populations and authorities for comprehensive guidance became so overwhelming, State donors eventually impelled Displacement Solutions to draft principles. As a result, the Peninsula Principles on Climate Displacement within States (2013) became the centre of this research.

The Peninsula Principles are the embodiment of a transnational governance innovation by non-State actors resulting from a growing sense of urgency around

[1] K. Hassine, Regularizing Property Rights in Kosovo and Elsewhere, 2010.

[2] K. Hassine, S. Leckie, The United Nations Principles on Housing and Property Restitution for Refugees and Displaced Persons: A Commentary, 2016.

climate displacement. They demonstrate today's increasing permeability of State-driven processes for non-State actors.

It is their reference to reality that gives nongovernmental, grassroots and civil society organizations like YPSA an irreplaceable role in global affairs. They are at the forefront of issues and crises and are thus best placed to inform the international debate and activity, and even to share their conclusions and actionable recommendations. This happens regularly in a timely and comprehensive manner, if these actors are being heard – a prerequisite for the sense of urgency to arise.

The 1992 United Nations Conference on Environment and Development in Rio de Janeiro illustrates this change in global affairs. It marked a turning point for civil society's role in intergovernmental processes with an unprecedented level of grassroots engagement. This evidenced – as United Nations Secretary-General Kofi Annan later acknowledged – that "[t]he nature of diplomacy [...] is changing everywhere to take in civil society. Traditionally, diplomacy has been an activity conducted exclusively by State actors and a subject debated exclusively by paid experts. In the United Nations a few decades ago, the Governments of Member States were virtually the sole players in the international process; non-governmental organizations (NGOs) were seen as supporters, allies, and mobilizers of public opinion in favour of the goals and values of the United Nations Charter."[3] Since then, non-State actors have continued their unprecedented move to center stage in world affairs, and today exert their influence in every single aspect of international policy-making.

3 SG/SM/6638, UN Press Release of July 14, 1998.

# 1

# Research Framework

> Climate change is an unprecedented and growing threat.
> The arguments for action are clear.
> So are the immense opportunities for peace and prosperity
> if we act quickly and decisively.
>
> United Nations Secretary-General António Guterres,
> "Climate Action: Mobilizing the World"

## I INTRODUCTION

### 1 *Background*

In December 2015, the world looked toward Paris. For more than twenty years, the international community had been trying to reach a universal, binding agreement on climate change. Paris was the international community's litmus test: States would either reach consensus on a multilateral instrument or it would be the end of the United Nations' efforts to forge global action on climate change. After two weeks of frantic negotiations, there was a historic breakthrough and the Paris Agreement was eventually adopted.

The Paris Agreement is hailed as a turning point. States, including developing nations, committed to reducing their greenhouse gas emissions. Developed countries also pledged to assist developing States, including through financial support, in their efforts to reduce emissions and to provide aid to those countries affected by climate-related disasters.

Beyond that, Paris can also be considered a turning point for human rights. It is the first climate change treaty that contains a reference to human rights and the need to respect human rights obligations when addressing climate change. By that, Paris paves the way for a rights-based approach to responses to climate change challenges. Similarly, the Paris Agreement contains a central reference to climate migrants in recognition of the fact that climate displacement may be one of the greatest challenges of the century.

And yet, Paris was a deplorable outcome for those affected by climate displacement. The provisions contained in the draft agreement on a climate change displacement coordination facility to help coordinate efforts to address the displacement of people as a result of the extreme impacts of climate change, notably through organized migration and planned relocation as well as compensation to those fleeing the adverse impact of climate change, did not meet the agreement of the negotiating parties and were deleted from the final Agreement.

This is even more disappointing since climate displacement is not a distant phenomenon, but is already happening in many countries around the globe. The envisaged climate change displacement coordination facility was expected to address the existing legal and institutional vacuum and provide guidance to States and assistance to affected communities.[1]

Mindful of the normative, policy and operational gap as a result of firsthand experience working with affected communities, Displacement Solutions, an international NGO working in the area of climate displacement, spearheaded the efforts to consolidate existing norms regarding climate displacement and procedures for relocation. As a result, in 2013, a drafting committee of experts and practitioners gathered to elaborate and adopt a set of principles known as the Peninsula Principles on Climate Displacement within States. The present study will examine whether in the absence of an internationally agreed framework, these principles provide an adequate response to handling climate displacement.

## 2 *Rationale and Research Question*

It is projected that climate change will lead to unprecedented numbers of displaced persons around the world in the decades ahead, even though this linkage has only recently been recognized. Most of those who will eventually have to leave their places of habitual residence will be remaining within their countries, while some may also cross international borders. So far, the international community has failed to provide adequate responses for the handling of climate displacement and for addressing the protection challenges of the affected individuals.

The emerging policy, operational and protection gaps need urgently to be addressed at both the international and national levels. While cross-border displacement may be covered, to some extent, by existing international protection frameworks and institutional arrangements, such as the refugee protection regime, there is no comprehensive response for many other cross-border movements as a result of climate change. Closing this gap at the international level may require setting into

1 FCCC/ADP/2015/L.6, December 5, 2015, Art. 5, para. 3: "A climate change displacement coordination facility shall be established under the [WIM] [Conference of the Parties serving as the meeting of the Parties to the Agreement (CMA)] to help coordinate efforts to address climate change induced displacement, migration and planned relocation." For further details, see J. Wentz, M. Burger, Designing a Climate Change Displacement Coordination Facility: Key Issues for COP 21, 2015.

motion an intergovernmental process to provide protection solutions for persons displaced across international borders. A recent attempt, the Nansen Initiative, which has been made at the international level to devise a suitable response in this regard, will also briefly be discussed in the context of this research in terms of its relevance to domestic climate displacement, even though cross-border displacement will not constitute the main focus of this study. This study will rather look at majority populations affected by environmental degradation and disasters that will remain within the boundaries of their own countries and will thus represent the core of the issue.

In spite of the recognition that climate displacement will be internal for most of its part,[2] few States have by now developed policies or established institutions to provide support to climate displaced persons in a comprehensive, all-encompassing and sustainable manner. This lack of action, which has multiple roots, including capacity gaps, is paralleled and exacerbated by the absence of concrete guidance and action-oriented recommendations at the international level. To address the challenges arising at the domestic level, States require guidance regarding their protection obligations toward individuals affected by climate change in order to conceptualize policies and design programs to assist those displaced.

This diagnosis, which emerged further to a bottom-up and broad-based consultative process, led a group of experts in August 2013 to elaborate the Peninsula Principles. The Peninsula Principles claims to be the first policy document that provides responses to the challenges of internal climate displacement.

A set of interrogations arise from this claim, the overall question being: is this nonbinding framework of the Peninsula Principles adequate to effectively address the policy, protection and operational challenges resulting from climate change displacement?

Before turning to an analysis of the Peninsula Principles themselves, the Principles' underlying assumptions (i.e. that climate change entails displacement and impacts human rights) will need to be examined. The basic set of questions is: Does climate change lead to displacement, and if so, what are the displacement scenarios and patterns? Is there a nexus between climate change and human rights? Is there a policy, protection and operational gap that needs to be filled? Or, in other words, what other frameworks and policy options exist?

## 3 Methodology

This research is situated at the intersection of political science and related disciplines such as international relations and public international law as well as human rights law.

[2] Government Office for Science, Foresight: Migration and Global Environmental Change, Final Project Report, 2011, p. 11; S. C. McAnaney, Sinking Islands? Formulating a Realistic Solution to Climate Change Displacement, 2012, p. 1178 *et subseq*.

It examines information collected through various means, including through discussions with experts and other stakeholders as well as exchanges of views at international conferences and consultations, such as the 2015 LANDac Annual International Conference in Utrecht,[3] the 2014 Sanremo consultation on Planned Relocation, Disasters and Climate Change,[4] and the 2015 Displacement Solutions Roundtable on Land Access for Climate Displaced Persons and Communities held in Geneva,[5] with a view to corroborate the findings and reduce potential bias of not only textual sources but also those that could result from the author's own involvement in the final drafting stage of the Peninsula Principles.

The analysis draws on a broad array of technical and legal primary and secondary documents, including an extensive body of literature from various disciplines. When it comes to the normative aspects of the handling of climate displacement, which cut across this study, the research uses established primary and secondary sources of international law in accordance with the authoritative enumeration contained in Article 38 of the Statute of the International Court of Justice.[6]

The initial part of the analysis, which frames the issue by looking at displacement scenarios, projections and patterns, draws on scientific climate change reports and related research from climate and social sciences, notably the writings of recognized academics and practitioners. The study then evaluates the various outcomes of the COP of the UNFCCC and other related mechanisms to analyze how displacement was progressively recognized within the climate change framework.[7] The body of human rights law and reports of human rights mechanisms are reviewed when examining the link between human rights and climate change in order to bring to the fore the diverging views relating to the recognition of this nexus.

The study then assesses the protection challenges to establish benchmarks for the evaluation of the Peninsula Principles and appraises existing policy options, basing itself on existing soft law and policy documents in addition to academic literature, which is available to a limited extent on more recent initiatives, including the Peninsula Principles themselves.

3 LANDac Conference, July 9–10, 2015, Session: Climate Change-Induced Displacement and Migration, 2015.

4 UNHCR/Brookings/Georgetown University, Planned Relocation, Disasters and Climate Change: Consolidating Good Practices and Preparing for the Future, Background Document, 2014; UNHCR, Planned Relocation, Disasters and Climate Change: Consolidating Good Practices and Preparing for the Future, Report, 2014.

5 Displacement Solutions, Land Access for Climate Displaced Persons and Communities, Roundtable hosted by Displacement Solutions and The Global Migration Centre of the Geneva Graduate Institute of International and Development Studies, Meeting Report, 2016; see also Displacement Solutions, Climate Displacement and Human Rights: Where to Go from Here?, Displacement Solutions Expert Roundtable Discussion, 2016.

6 K. Oellers-Frahm, Art. 92, G. I., para. 101 *et subseq.*, in: B. Simma *et al.*, The Charter of the United Nations, Vol. II, 2012.

7 See http://unfccc.int/bodies/body/6383.php (accessed May 22, 2019) for an updated overview.

Turning to the assessment of the Peninsula Principles, rather than analyzing them on a principle-by-principle basis, the study follows a thematic cluster approach and offers commentary on the significance of the Principles' value from a policy, operational and legal perspective, drawing on *travaux préparatoires* and a variety of other primary and secondary sources across several disciplines, thereby critically assessing whether the protection challenges are effectively addressed and how the Principles need to be operationalized to contribute to this effect.

## 4 *Scope*

This research focuses on how to respond adequately to internal displacement situations occurring because of the effects of climate change through the application of the Peninsula Principles and on how to ensure protection of the rights of those affected, including individual persons or communities facing climate displacement. It does not conduct an analysis into empirical evidence relating to the issue of climate change itself. Such scientific evidence can be found notably in the reports of the IPCC, which are issued on a regular basis.[8]

This study moreover addresses climate displacement within States (i.e. displaced persons who do not cross an internationally recognized border as a result of the adverse effects of climate change). Cross-border displacement will not be the focus of this research. Also not within the scope of this research is displacement resulting from industrial accidents as well as development-induced displacement and resettlement (i.e. displacement due to environmental degradation that occurs as a result of the implementation of development projects).

## 5 *Terminology and Discourse*

There is no consensus on how to designate people on the move as a result of climate change or the phenomenon itself. Terms employed try to link existing concepts either with progressive environmental degradation, with the occurrence of disasters or more specifically with the climate change phenomenon and its consequences and range from *migrants* to *displaced persons* and *refugees*. While some of the terms are well defined – or at least circumscribed – in international law, there are no universally agreed definitions of the compound terms.[9]

An additional layer of complexity is added by the fact that population movements are rarely monocausal. An individual may for a variety of reasons decide or be compelled to leave their habitual place of residence, one of which may be the degradation of the environment and the loss of livelihood as a result of climate change. There are intersections with other push factors for migration, such as the

8 Available at www.ipcc.ch/publications_and_data/publications_and_data_reports.htm (accessed March 4, 2019).

9 K. Warner, Global Environmental Change and Migration: Government Challenges, 2010, p. 403.

socioeconomic environment, particularly as climate change disproportionately affects the most vulnerable segments of society, and which of these is dominant in the decision may be difficult to determine.[10]

Migration theory places the decision to migrate on a volitional continuum, which ranges from entirely voluntary migration, where the individual has total control over their relocation, to involuntary or forced migration, characterized by the total absence of control.[11] In other words, proactive and reactive migration.[12] Distinctions are also made with regard to the timing of the migration decision, which seems to be particularly relevant for the case of climate displaced persons. Egon Kunz notably differentiates between *acute refugees*, fleeing a crisis, and those who leave before the actual crisis occurs (i.e. *anticipatory refugees*).[13]

For the reasons outlined, it is important to examine in more depth the terminology in use in the current climate change/human mobility discourse, with a view to identifying the exact meaning and legal connotations of the various terms, which are often employed interchangeably and indiscriminately, and to determine which of them best captures the complexities of population movements resulting from climate change.

### a Climate Change, Environmental Degradation and Disasters

Climate change, according to Article 1 of the UNFCCC, is "a change of climate which is attributed directly or indirectly to human activity that alters the composition of the global atmosphere and which is in addition to natural climate variability observed over comparable time periods." Whereas the Convention distinguishes between anthropogenic climate change (i.e. the changes in the atmospheric composition that can be attributed directly or indirectly to human activities) and climate variability (i.e. the natural changes of the climate over time), the IPCC does not draw that distinction as it defines climate change as any change in climate,[14]

[10] C. Gibb, J. Ford, Should the United Nations Framework Convention on Climate Change Recognize Climate Migrants?, 2012, p. 2; J. McAdam, Swimming against the Tide: Why a Climate Change Displacement Treaty Is Not the Answer, 2011.

[11] E. Kunz, The Refugees in Flight: Kinetic Models and Forms of Displacement, 1973, pp. 125–146; A. H. Richmond, Sociological Theories of International Migration: The Case of Refugees, 1988, pp. 7–25; A. H. Richmond, Reactive Migration: Sociological Perspectives on Refugee Movements, 1993, pp. 7–24; E. Fussell, Space, Time and Volition: Dimensions of Migration Theory, 2012, p. 38 *et subseq.*; D. C. Bates, Environmental Refugees? Classifying Human Migrations Caused by Environmental Change, 2002, p. 467 *et subseq.*; Greenpeace, Klimaflüchtlinge, Die verleugnete Katastrophe, 2007, p. 11.

[12] A. H. Richmond, Sociological Theories of International Migration: The Case of Refugees, 1988, pp. 7–25; A. H. Richmond, Reactive Migration: Sociological Perspectives on Refugee Movements, 1993, pp. 7–24.

[13] E. Kunz, The Refugees in Flight: Kinetic Models and Forms of Displacement, 1973, pp. 125–146.

[14] It is important to note that there are inconsistencies in the IPCC's use of the climate change definition. Whereas the working group on science developed the mentioned definition, the working group on mitigation bases its analyses on the UNFCCC definition and the working group on impacts,

irrespective of whether it is linked to the natural variability or human activity.[15] The definition contained in the UNFCCC is therefore more limited in scope, which is congruent with the main purpose of the Convention, which is primarily concerned with States' action, seeking to commit them to reduce their greenhouse gas emissions. This aspect of climate change, which is the result of human interference and is within the realm of States, constitutes the UNFCCC's main focus that consequently determines its definition of the climate change phenomenon. The objective of the IPCC in turn "is to assess on a comprehensive, objective, open and transparent basis the scientific, technical and socio-economic information relevant to understanding the scientific basis of risk of human-induced climate change, its potential impacts and options for adaptation and mitigation."[16] The scope of its mandate accordingly spans beyond changes resulting from human activity.

The definitions' distinct scope is obviously important, as this distinction will determine the latitude of protection accorded to people on the move. The mismatch between these definitions also has the potential to seriously impact the response of the international community to the challenges of climate change, as the political process (i.e. the intergovernmental process) follows the UNFCCC parameters and is hence based on its more limited definition.

From a rights-based perspective, focusing on certain aspects of climate change only, namely the anthropogenic part, would not be practicable. Consequently, the Peninsula Principles follow the IPCC definition. Principle 2(a) defines climate change as "the alteration in the composition of the global atmosphere that is in addition to natural variability over comparable time periods."

There is a variety of operational definitions of disaster in use within the United Nations, which demonstrates again the difficulties to formulate universally acceptable definitions that would satisfy all actors, who, depending on their mandate, may have a different focus and different needs.[17] The UNDP defines disasters as "occurrences when hazards, whether natural or human-made, interact with vulnerable populations,"[18] while the OCHA includes within the scope of its definition, the impact of disasters on society. According to the OCHA a disaster is "a serious disruption of the functioning of a society, causing widespread human, material, or environmental losses which exceed the ability of affected society to cope using its own resources. Disasters are often classified according to their speed of onset

adaptation and vulnerability seems to use both. See R. A. Pielke Jr., Misdefining "Climate Change": Consequences for Science and Action, 2005, p. 555.

15 A. Verbruggen, W. Moomaw, J. Nyboer, Annex I: Glossary, Acronyms, Chemical Symbols and Prefixes. In IPCC Special Report on Renewable Energy Sources and Climate Change Mitigation, 2011.

16 See IPCC, Principles Governing IPCC Work, para. 2.

17 For an academic discussion of disaster definitions, see, e.g., R. W. Perry, What Is a Disaster?, 2007, pp. 1–15; E. Ferris, The Politics of Protection, The Limits of Humanitarian Action, 2011, p. 200 *et seq.*; see also CEDAW/C/GC/35 of July 14, 2017, for the gender dimension.

18 IASC, Definition of Complex Emergencies, 1994.

(sudden or slow), or according to their cause (natural or man-made)."[19] The most authoritative of definitions is the one used by the UNISDR, which does not only consider the impact on society as a whole, but is also concerned with the community level. It also reduces the threshold of the losses that have to be suffered in order to qualify as a disaster by leaving out *widespread*. The UNISDR defines a disaster as follows: "A serious disruption of the functioning of a community or a society at any scale due to hazardous events interacting with conditions of exposure, vulnerability and capacity, leading to one or more of the following: human, material, economic and environmental losses and impacts."[20] It provides its own interpretation of the definition, noting: "The effect of the disaster can be immediate and localized, but is often widespread and could last for a long period of time. The effect may test or exceed the capacity of a community or society to cope using its own resources, and therefore may require assistance from external sources, which could include neighbouring jurisdictions, or those at the national or international levels."[21]

*Environmental degradation* is another term used to qualify population displacement. Environmental phenomena are not necessarily linked to climate change and are much broader – and perhaps even too broad – as they could also include industrial accidents and alike. On the other hand, precisely because of the challenges in affirming the causal relation between sudden and particularly slow-onset disasters and displacement, this may be the preferred solution.

### b Human Mobility

*Human mobility* is the terminology used by the IOM to describe migration in general terms. This concept is all-encompassing "ranging from totally voluntary migration [. . .] to totally forced migration,"[22] reflecting the volitional continuum of migration theory.[23] The Nansen Initiative, by contrast, which is concerned with cross-border movement as a result of climate change, uses the Cancún Climate Change Adaptation Framework (CAF) as a point of departure and therefore also includes *relocation* within the scope of its definition of human mobility. Paragraph 14(f) of the CAF refers to displacement, migration and planned relocation, which

19 *Ibid.*

20 UNISDR, Terminology on Disaster Risk Reduction, 2009.

21 *Ibid.*

22 IOM, Migration, Environment and Climate Change: Evidence for Policy (MECLEP), Glossary, July 2014.

23 E. Kunz, The Refugees in Flight: Kinetic Models and Forms of Displacement, 1973, pp. 125–146; A. H. Richmond, Sociological Theories of International Migration: The Case of Refugees, 1988, p. 7 *et seq.*; A. H. Richmond, Reactive Migration: Sociological Perspectives on Refugee Movements, 1993, pp. 7–24; E. Fussell, Space, Time and Volition: Dimensions of Migration Theory, 2012, p. 38 *et subseq.*; D. C. Bates, Environmental Refugees? Classifying Human Migrations Caused by Environmental Change, 2002, p. 467 *et subseq.*; Greenpeace, Klimaflüchtlinge, Die verleugnete Katastrophe, 2007, p. 11.

designates – according to the Nansen Initiative – "a planned process of settling persons or groups of persons to a new location,"[24] and which may be either voluntary or involuntary in nature.

### c Climate Refugees

*Climate refugee* is not a recognized term under international law. A refugee under Article 1 of the 1951 Refugee Convention, as amended by the 1967 Protocol, is:

> A person who owing to a well-founded fear of being persecuted for reasons of race, religion, nationality, membership of a particular social group or political opinion, is outside the country of his nationality and is unable or, owing to such fear, is unwilling to avail himself of the protection of that country; or who, not having a nationality and being outside the country of his former habitual residence as a result of such events, is unable or, owing to such fear, is unwilling to return to it.[25]

The terms *climate* or *environmental refugee* therefore are misleading, as they imply that the individual is fleeing war or persecution and can therefore claim a certain protection under the refugee regime.[26] They also only cover those individuals who have crossed international borders and not those who stay within their country.

Beyond the elements contained in the 1951 Refugee Convention, the 1969 Organization of African Unity Convention governing the specific aspects of refugee matters in Africa, in its Article 1, defines a refugee as any person compelled to leave their place of habitual residence in order to seek refuge in another place outside their country of origin or nationality, "owing to external aggression, occupation, foreign domination or events seriously disturbing public order in either part or the whole of [their] country of origin or nationality." While this definition is again limited to cross-border movement, it is broader than the *classical* refugee definition embracing the notion of *events seriously disturbing public order*, which could occur as a result of environmental degradation and loss of livelihood due to the adverse impact of climate change. The threshold for events seriously disturbing public order being lower than that of conflict, this definition could cover many more individuals who flee the impact of climate change, though it will still not extend to monocausal motivations.[27]

24 Nansen Initiative, Agenda for the Protection of Cross-Border Displaced Persons in the Context of Disasters and Climate Change, 2015, p. 17; see also UNHCR, Planned Relocation, Disasters and Climate Change: Consolidating Good Practices and Preparing for the Future, Report, 2014.

25 For the full text of the 1951 Refugee Convention, see www.unhcr.org/protect/PROTECTION/3b66c2aa10.pdf (accessed April 18, 2019).

26 B. L. Burson: [2013] NZIPT 800413, para. 51.

27 African Union Convention Governing the Specific Aspects of Refugee Problems in Africa, adopted on September 10, 1969, by the Assembly of Heads of State and Government. CAB/LEG/24.3. It entered into force on June 20, 1974.

**The Kampala Convention**

The Kampala Convention, which came into force on December 6, 2012, is the first regional instrument that provides legally binding protection for the rights and well-being of those forced to flee inside their home countries due to conflict, violence, natural disasters and human rights abuses.

More specifically, the Convention:

- promotes and strengthens regional and national measures to prevent and eliminate the root causes of internal displacement and provide for solutions;
- establishes a legal framework for prevention, protection and assistance to IDPs;
- establishes a legal framework for solidarity, cooperation and promotion of durable solutions between State parties;
- outlines the obligations and responsibilities of Member States;
- outlines the obligations, roles and responsibilities of armed groups, non-State actors and other relevant actors including civil society organizations and humanitarian agencies.

The Convention makes specific reference to climate change in Article 5(4):

> States parties shall take measures to protect and assist persons who have been internally displaced due to natural or human made disasters, including climate change.

See: Explanatory Note on the African Union Convention for the Protection and Assistance of Internally Displaced Persons in Africa (Kampala Convention), www.internal-displacement.org/sites/default/files /inline-files/200910-training-KC-Explanatory-Note-Eng.pdf

The UNHCR expressed serious reservations about the association of the term *refugee* with situations not currently covered by the 1951 Refugee Convention. There are concerns that this could undermine the international refugee protection regime. The UNHCR recognizes that climate change and environmental degradation may be among the root causes pushing persons to cross international borders. This is not enough, however, under applicable international law to qualify them as refugees. While some of the migrants may fall within the scope of the international protection regime, there is a protection gap with regard to those who cannot on this basis claim refugee status under international law, even though they are in actual need of assistance.[28]

[28] UNHCR, Climate Change, Natural Disasters and Human Displacement: A UNHCR Perspective, 2009; see also W. Kälin, Displacement Caused by the Effects of Climate Change: Who Will Be Affected and What Are the Gaps in the Normative Framework for Their Protection?, 2008; O. Brown, Migration and Climate Change, 2008, p. 36 *et seq.*; D. Kniveton *et al.*, Climate Change and Migration: Improving

Those in favor of the term *climate* or *environmental refugee* argue that this would better capture the involuntary or forced nature of the individual's movement as opposed to migration.[29] There are voices calling for an amendment of the Refugee Convention to include people who crossed borders as a result of climate change. The UNHCR opposes such proposals as this would require renegotiating the Refugee Convention, with the risk of undermining existing protection standards.[30]

### d Climate Migrants

*Migration* denotes "[a] process of moving, either across an international border, or within a State. It is a population movement, encompassing any kind of movement of people, whatever its length, composition and causes; it includes migration of refugees, displaced persons, uprooted people, and economic migrants." It is therefore an all-encompassing term referring to various forms of movement including voluntary and involuntary ones.[31]

It is important to note, however, that unlike the term *refugee*, the term *migrant* is not defined under international law and that no particular protection regime is therefore applicable. Taking the aforementioned into account, the IOM developed a working definition of the term *climate migration*, which encompasses voluntary as well as forced movements. Accordingly, climate migration is defined as "[t]he movement, within a State or across an international border, of a person or groups of persons, who are obliged to leave their habitual place of residence, or choose to do so, either temporarily or permanently, predominantly for reasons of sudden or progressive change in their environment due to climate change."[32] While this definition is very broad and places the emphasis on the movement itself, it is interesting to note that the IOM in its more general migrant definition, which again has no normative value, states that the term "is usually understood to cover all cases where the decision to migrate is taken freely by the individual concerned for reasons of 'personal convenience' and without intervention of an external compelling factor."[33] Such interpretation, limiting migration to voluntary movements, would, as a matter of fact, exclude a number of individuals from the scope of the

Methodologies to Estimate Flows, 2008, p. 31 *et seq.*; B. Docherty, T. Giannini, Confronting a Rising Tide: A Proposal for a Convention on Climate Change Refugees, 2009, p. 358; D. Hodgkinson *et al.*, "The Hour When the Ship Comes In": A Convention for Persons Displaced by Climate Change, 2010; A. Williams, Turning the Tide: Recognizing Climate Change Refugees, 2008.

29 *Cf.* J. Haslinger, Tsunami: A Report from Phi Phi Island, 2007, p. 119, in which the tsunami victims are referred to as refugees.

30 UNHCR, Climate Change, Natural Disasters and Human Displacement: A UNHCR Perspective, 2009.

31 IOM, International Migration Law, Glossary on Migration, 2004, p. 41.

32 IOM, Migration, Environment and Climate Change: Evidence for Policy (MECLEP), Glossary, 2014, p. 13.

33 IOM, International Migration Law, Glossary on Migration, 2004, p. 41.

definition as in many cases, if not most, the environmental degradation may be such that it actually becomes a compelling factor.

The multicausality of push factors as well as the spectrum of migration categories are well captured by the definition used by the IPCC in its fifth assessment report, which states:

> Human migration involves movement over a significant distance and duration. Environmental migration refers to human migration where environmental risks or environmental change plays a significant role in influencing the migration decision and destination. Migration may involve distinct categories such as direct, involuntary, and temporary displacement due to weather-related disasters; voluntary relocation as settlements and economies become less viable; or planned resettlement encouraged by government actions or incentives. All migration decisions are multicausal, and hence it is not meaningful to describe any migrant flow as being solely for environmental reasons.[34]

Nevertheless, this definition does not reflect the degree of compulsion involved when leaving the place of residence because of environmental degradation or disasters, including slow-onset climate events. It also does not provide any indication of the applicable protection regime under international law.

It may be relatively easy to establish the forced character of climate migration in the case of natural disasters and even those that are imminent,[35] but in the case of environmental degradation, particularly at its early stages, affirming the forced nature of the movement may be more contentious. Another element of ambiguity linked to the use of the migration terminology in this context is how to qualify adaptation measures, such as voluntary relocation or resettlement. Some authors argue that these are forms of voluntary migration even if they are triggered by environmental degradation and therefore express their support for the broader terminology of climate migrant as opposed to climate displaced person.[36]

### e Climate Displacement

The forced nature of movement is best reflected by the term *climate* or *environmentally displaced person*. According to the IOM, environmentally displaced persons are those "displaced within their country of habitual residence or who have crossed an international border and for whom environmental degradation, deterioration or destruction is a major cause of their displacement, although not necessarily the

34 IPCC, Climate Change 2014: Synthesis Report, 2014, p. 1765.

35 Brookings Institution, The Year That Shook the Rich: A Review of Natural Disasters in 2011, 2012, p. 5 *et seq.*

36 A. Suhrke, Environmental Degradation and Population Flows, 1994, p. 473 *et seq.*; G. Hugo, Environmental Concerns and International Migration, 1996, p. 105 *et seq.*; F. Renaud *et al.*, Control, Adapt or Flee: How to Face Environmental Migration?, 2007; O. Dun, F. Gemenne, Defining "Environmental Migration," 2008, p. 10 *et seq.*

sole one."[37] In line with this definition, the displacement may be internal or involve cross-border movements. It also takes account of the fact that decisions to leave the habitual place of residence are seldom monocausal and may involve environmental degradation and loss of livelihoods to varying degrees.

As climate change will primarily create internal displacement before it may eventually trigger people to move across borders, the definition of IDPs provided by the Guiding Principles on Internal Displacement, which reflect international human rights and humanitarian law, should be considered as well: "[P]ersons or groups of persons who have been forced or obliged to flee or to leave their homes or places of habitual residence, in particular as a result of or in order to avoid the effects of armed conflict, situations of generalized violence, violations of human rights or natural or human-made disasters, and who have not crossed an internationally recognized State border."[38] The element of compulsion (i.e. the idea of involuntary movement) is contained in this definition, while it focuses on disasters, either natural or anthropogenic ones. The element of coercion includes the threat to life and livelihood.

This definition drew on that contained in the 1992 analytical report of the then Secretary-General of ECOSOC, which the Representative of the Secretary-General on internally displaced persons, Walter Kälin, later adopted as a working definition and which he adapted while elaborating the Guiding Principles on Internal Displacement. In the report, IDPs were defined as "persons who have been forced to flee their homes suddenly or unexpectedly in large numbers, as a result of armed conflict, internal strife, systematic violations of human rights or natural or man-made disasters; and who are within the territory of their own country."[39] References to "suddenly or unexpectedly" and "large numbers" were removed to broaden the scope of application to situations of internal displacement irrespective of their scale. One of the difficulties with these displacement definitions is the question whether all movements within a State associated with climate change necessarily have to be considered forced or whether there are voluntary movements as well.[40]

The Peninsula Principles consider climate displacement to mean "the movement of people within a State due to the effects of climate change, including sudden and slow-onset environmental events and processes, occurring either alone or in combination with other factors," thus adapting the definition of the UN Guiding Principles to the specific situation of climate change, while not explicitly mentioning the degree of coercion associated with the displacement. Another element that is introduced by its definition of climate displaced persons is that it extends to "individuals, households or communities who are facing or experiencing climate

37 IOM, Migration, Environment and Climate Change: Evidence for Policy (MECLEP), Glossary, 2014.

38 E/CN.4/1998/53/Add.2.

39 E/CN.4/1992/23.

40 *Cf.* OCHA, Reference Guide, 2nd ed., 2011.

displacement," thereby including the threat within the scope of its application.[41] This seems to build on notions introduced by migration theory, namely the anticipatory character of displacement.[42] While there are concerns that the scope of this definition, which goes beyond the definition of the Guiding Principles, is too broad, thus not operational, there seems to be no justification for a distinction between the protection afforded to IDPs within the meaning of the definition of the Guiding Principles and other internally displaced.[43] The Peninsula Principles have also sought to address the issue of causality by deleting the reference to *climate change* and focusing instead on *climate displacement*. An earlier iteration of the Principles was in fact entitled Guiding Principles on Climate Change Displacement.[44] However, in the definition of climate displacement itself, the linkage to climate change reappears.

The Agenda for the Protection of Cross-Border Displaced Persons in the Context of Disasters and Climate Change, the so-called Nansen Initiative, followed a different path as it focuses on disaster-induced cross-border displacement. It encompasses "displacement in the context of disasters linked to hydro-meteorological and climatological hazards like flooding, tornadoes, cyclones, drought, salt water intrusion and glacial melting as well as geophysical hazards such as earthquakes, tsunamis or volcanic eruptions. It considers the effects of both sudden-onset and slow-onset hazards."[45] This approach therefore focuses on the effects of climate change (i.e. the cause of displacement) rather than on the nature of the disaster (i.e. whether it is or not human-made).

### f Other Definitional Attempts

Some authors engaged in definitional attempts that do not necessarily follow the delineation of established concepts, such as refugees, migrants or displaced persons, which are, if not legally defined, well-established in international affairs. The importance in these cases is not the term *environmental refugee*, as may be understood in international law, but rather the definition ascribed to it by the various authors, which tries to capture the reality of people on the move due to the adverse impact of climate change by reducing their motives and circumstances to those essential elements that can be generalized. As a result, the term *refugee* in this context is redefined and devoid of its original meaning and any legal consequences attached to it.

41 The Peninsula Principles on Climate Displacement within States, 2013, Principle 2.
42 E. Kunz, The Refugees in Flight: Kinetic Models and Forms of Displacement, 1973, pp. 125–146.
43 K. Koser, Gaps in IDP Protection, 2008, p. 17.
44 D. Hodgkinson, A Brief Overview of the Drafting of the Peninsula Principles, 2016, p. 33.
45 Nansen Initiative, Agenda for the Protection of Cross-Border Displaced Persons in the Context of Disasters and Climate Change, 2015.

The term *environmental refugee* was first used in the 1970s by Lester Brown of Worldwatch Institute and later propagated and popularized through a 1985 UNEP report drafted by Essam El-Hinnawi. Therein environmental refugees were defined as "[t]hose people who have been forced to leave their traditional habitat, temporarily or permanently, because of a marked environmental disruption (natural and/or triggered by people) that jeopardized their existence and/or seriously affected the quality of their life."[46]

In 1988, Jodi Jacobson noted that "[e]nvironmental refugees have become the single largest class of displaced persons in the world." She identified three broad categories of environmental refugees: "(1) those displaced temporarily because of a local disruption such as an earthquake or avalanche, (2) those who migrate because of environmental degradation that undermines their livelihood or threatens their health, or (3) those who resettle because land degradation has resulted in desertification or because of other permanent and untenable changes in their habitat."[47]

Diane Bates suggested the following definition of environmental refugees as, "people who migrate from their usual residence due to changes in their ambient non-human environment." She also provided a categorization of environmental refugees along the origin of the environmental disruption (i.e. whether it is natural or technological), the duration (i.e. acute or gradual) and whether migration was a planned outcome or not. She further distinguished three types of migration flows, prompted by unplanned outcomes and disruptions, namely disasters, expropriations and deterioration.[48]

Another definition of environmental refugees was proposed by Norman Myers in 2002. According to him, environmental refugees are

> people who can no longer gain a secure livelihood in their homelands because of drought, soil erosion, desertification, deforestation and other environmental problems, together with associated problems of population pressures and profound poverty. In their desperation, these people feel they have no alternative but to seek sanctuary elsewhere regardless of any hazardousness of their attempt. Not all of them have fled their countries, many being internally displaced. But all have abandoned their homelands on a semi-permanent if not permanent basis, with little hope of a foreseeable return.[49]

Frank Biermann and Ingrid Boas define climate refugees as "people who have to leave their habitats, immediately or in the near future, because of sudden or gradual alterations in their natural environment related to at least one of three impacts of

[46] E. El-Hinnawi, Environmental Refugees, 1985; see also International Institute for Environment and Development (IIED), Environmental Refugees, 1984; G. Kibreab, Environmental Causes and Impact of Refugee Movements: A Critique of the Current Debate, 1997, p. 20 *et seq.*

[47] J. L. Jacobson, Environmental Refugees. A Yardstick of Habitability, 1988, p. 37 *et subseq.*

[48] D. C. Bates, Environmental Refugees? Classifying Human Migrations Caused by Environmental Change, 2002, p. 465 *et seq.*

[49] N. Myers, Environmental Refugees: A Growing Phenomenon of the 21st Century, 2002, p. 609 *et seq.*

climate change: sea-level rise, extreme weather events, and drought and water scarcity."[50] Fabrice Renaud *et al.* finally distinguish environmentally motivated migrants, environmentally forced migrants and environmental refugees.[51]

Most of these definitions combine the elements of coercion with the environmental triggers and the impact on the lives of people.

## g Concluding Summary

In summary, *climate* and *environmental* both seek to qualify the causes of the human movement, but while *climate* directly refers to the adverse impact of climate change, encompassing human-made and natural environmental degradation in accordance with the IPCC's definition, *environmental* is broader and its scope may extend to disaster or environmental degradation caused by phenomena other than climate change, for instance industrial accidents. *Climate*, however, implies that a clear link to climate change can be established. Whereas it may not be easy in all instances to determine this causal relation, the use of *environmental* in turn would broaden the scope much beyond the intended focus and therefore is less preferable in the instant case. As for the use of *disaster*, while there may be a clearer nexus between the event and the movement of people, particularly in the case of sudden and rapid-onset disasters, it includes many types of hazards, such as biological, environmental, geological, hydrometeorological and technological processes and phenomena, which are not all the result of climate change.[52]

The term *climate migration* encompasses all types of movements, voluntary and involuntary, associated with the effects of climate change, and allows to reflect that migration decisions are situated on a volitional continuum. The *climate refugee* terminology, by contrast, whereby *refugee* has a clearly defined legal scope, captures the assumed coercive character of the movement as a result of climate change, but seems misleading in those cases at least where the element of persecution is absent and therefore does not lead to the application of the international refugee protection regime. It also bears the risk to undermine this regime if the terminology were to be renegotiated to broaden its scope to a new category of cases and is therefore being rejected by the UNHCR. As for the designation *climate displacement*, this may hence be the most appropriate one, as it also implies a degree of coercion, distinguishing it from the very unspecific migration terminology. This research will base itself on the definition of climate displacement as understood by the Peninsula Principles, as the subject matter of the analysis.

50 F. Biermann, I. Boas, Preparing for a Warmer World: Towards a Global Governance System to Protect Climate Refugees, 2010, p. 67.

51 F. Renaud *et al.*, Control, Adapt or Flee: How to Face Environmental Migration?, 2007, p. 29 *et seq.*

52 See A/CONF.224/CRP.1.

## 6 Overview of Research Streams

Climate displacement started entering scientific research sporadically in the 1970s and 1980s. With displacement being one of the palpable consequences of climate change, scientific debate on this particular aspect was geared toward raising more general awareness about the phenomenon of climate change and the potential seriousness of its impact. Early reports focusing on the link between climate change and migration projected population movements of an unprecedented magnitude, thereby trying to alarm the international community and instigate related action.[53] This led to further research in the area of international relations focusing on the adverse impact of climate displacement, including with regard to the particular security aspects of climate migration, in particular the potential to destabilize States, and trigger social unrest and conflicts.[54] The focus of this early research on environmental migrants being uprooted as a result of climate change exhibited diverging views in natural science, which affirmed the causal relationship between migration and environmental degradation, and social science, which took a more nuanced approach rejecting monocausal explanations of migration.[55]

Studies on climate migration and displacement have since become more frequent, but the issue only became a focus of applied research since the IPCC's first assessment report in 1990. This intensified research and also led to a more systematic analysis of the climate change–migration nexus in the following decade, particularly through case studies and empirical studies.[56] The body of literature grew significantly with the beginning of the new century, adding a conceptual and methodological layer to the analysis of the climate displacement phenomenon and bringing about in-depth studies of the political implications of climate change and its effects

53 L. R. Brown, World Population Trends: Signs of Hope, Signs of Stress, 1976; E. El-Hinnawi, Environmental Refugees, 1985; N. Myers, Environmental Refugees in a Globally Warmed World, 1993, p. 752 *et seq.*; J. L. Jacobson, Environmental Refugees: A Yardstick for Habitability, 1988.

54 R. H. Ullman, Redefining Security, 1983, p. 129 *et seq.*; T. F. Homer-Dixon, On the Threshold: Environmental Changes as Causes of Acute Conflict, 1991, p. 76 *et seq.* See also J. Barnett, Security and Climate Change, 2001; O. Brown, Weather of Mass Destruction? The Rise of Climate Change as a "New" Security Issue, 2007; W. A. V. Clark, Environmentally Induced Migration and Conflict, 2006; N. P. Gleditsch, R. Nordås and I. Salehyan, Climate Change and Conflict: The Migration Link, 2007; F. Sindico, Climate Change: A Security (Council) Issue? 2007, p. 29 *et seq.*; see also R. McLeman, Climate Change, Migration and Critical International Security Considerations, 2011, p. 25 *et seq.*; see also H. Welzer, Klimakrieg, Wofür im 21. Jahrhundert getötet wird, 2010; OCHA, Understanding the Climate–Conflict Nexus from a Humanitarian Perspective: A New Quantitative Approach, 2016.

55 R. Black, Environmental Refugees: Myth or Reality?, 2001; S. Castles, Environmental Change and Forced Migration: Making Sense of the Debate, 2002.

56 C. Raleigh, L. Jordan, I. Salehyan, Assessing the Impact of Climate Change on Migration and Conflict, 2008; D. K. Bardsley, G. J. Hugo, Migration and Climate Change: Examining Thresholds of Change to Guide Effective Adaptation Decision-Making, 2010, p. 238 *et seq.*; E. Piguet, A. Pécoud, P. de Guchteneire, Migration and Climate Change: An Overview, 2011, p. 1 *et seq.*; C. Gray, R. Bilsborrow, Environmental Influences on Human Migration in Rural Ecuador, 2013, p. 1 *et seq.*

on population movements. Other more recent research highlighted the knowledge gaps and aimed at providing a more refined analysis of how the climatic changes, including slow-onset disasters, will affect populations and lead to migration/displacement.[57] Also, the IPCC's fifth assessment report dedicates a subchapter to this topic.[58]

## II CLIMATE DISPLACEMENT SCENARIOS

### 1 *Projections*

The linkages between climate change and displacement are complex and cannot entirely be predicted.[59] Projections and estimates of numbers of people who will be forced to move as a result of climate change therefore remain controversial and diverge considerably depending on the specific circumstances and the parameters used.[60] Another reason for low confidence in quantitative projections of migration/displacement patterns is the complex, multicausal nature of migration.[61]

The IPCC in its report for instance estimated that by 2050, 150 million people could be displaced by climate change-induced phenomena, such as desertification, water scarcity, floods and storms,[62] whereas more recent and frequently cited estimates project that 200 million will be forcibly displaced by the year 2050, losing their homes, land and property.[63] There is no consensus about the number of

57 See inter alia G. Hugo, Migration, Development and Environment, 2008; D. Kniveton *et al.*, Climate Change and Migration: Improving Methodologies to Estimate Flows, 2008; E. Piguet, Climate Change and Forced Migration, 2008; J. Jäger *et al.*, Environmental Change and Forced Migration Scenarios, Synthesis Report, 2009; J. Morrissey, Environmental Change and Forced Migration: A State of the Art Review, 2009; C. Tacoli, Crisis or Adaptation? Migration and Climate Change in a Context of High Mobility, 2009, p. 513 *et seq.*; O. Brown, Migration and Climate Change, 2008; S. Perch-Nielsen, M. B. Bättig, D. Imboden, Exploring the Link between Climate Change and Migration, 2008, p. 375 *et seq.*; G. Jonsson, The Environmental Factor in Migration Dynamics: A Review of African Case Studies, 2010; S. F. Martin, Climate Change, Migration and Governance, 2010, p. 397 *et seq.*; V. Kolmannskog, Climate Change, Human Mobility, and Protection: Initial Evidence from Africa, 2010, p. 103 *et seq.*; F. Biermann, I. Boas, Preparing for a Warmer World: Towards a Global Governance System to Protect Climate Refugees, 2010, p. 60 *et seq.*

58 IPCC, Climate Change 2014: Impacts, Adaptation, and Vulnerability, Summary for Policymakers, p. 20.

59 See E. Piguet, A. Pécoud, P. de Guchteneire, Migration and Climate Change: An Overview, 2011, p. 12 *et subseq.*; O. Brown, Migration and Climate Change, 2008, p. 21 *et seq.*; OCHA, Internal Displacement Monitoring Centre (IDMC), Norwegian Refugee Council (NRC), Monitoring Disaster Displacement in the Context of Climate Change, 2009, annex 1.

60 O. Dun, F. Gemenne, Defining "Environmental Migration," 2008, p. 10 *et seq.*

61 IPCC, Climate Change 2014: Impacts, Adaptation, and Vulnerability, Summary for Policymakers, p. 20.

62 IPCC, First Assessment Report, 1990.

63 IOM, Migration and Climate Change, 2008, p. 11 *et subseq.*; N. Myers, J. Kent, Environmental Exodus: An Emergent Crisis in the Global Arena, 1995; N. Myers, Environmental Refugees: An Emergent Security Issue, 2005; N. Stern, The Economics of Climate Change: The Stern Review, 2006, p. 3; see also K. Warner, Climate Change Induced Displacement: Adaptation Policy in the Context of the UNFCCC Climate Negotiations, 2011, p. 2; see also O. Brown, The Numbers Game, 2008, p. 8 *et subseq.*; K. Warner, F. Laczko, A Global Research Agenda, 2008, p. 59 *et seq.*;

persons that are likely to be displaced, as this will also depend on the level of investment, planning and resources by governments.[64]

The difficulties climate change frameworks face in affirming that climate change may lead to displacement are also related to the inherent differences of social sciences, which would look at the potential impact of climate change on human mobility and climate change science itself. Obviously, the definition of who falls within the category of climate displaced will also significantly determine estimates.[65]

Irrespective of the varying estimates and other issues, such as causality/path-dependency, as well as resilience and adaptive capacity, views have by now mostly converged that climate change will lead to population displacements.[66] In addition to mitigation in order to limit the phenomenon of climate change and its impact to the extent possible, suitable adaptation strategies need to be devised.[67] Human mobility will be one of the many ways in which affected populations will be adapting to their changed environment, and displacement may occur in an unprecedented magnitude, raising significant challenges.[68]

Another important aspect, particularly when it comes to developing adequate normative and/or policy responses, is that most experts concur that climate displacement will be primarily internal.[69]

## 2 *Distributional Impact*

Another aspect of climate change that is often highlighted is its unequal distributional impact. Climate change will tend to affect developing countries in a disproportionate manner, thus further exacerbating poverty and unequal access to resources such as food and clean drinking water.[70] Again, there is a degree of uncertainty in relation to estimates, and climate change will not impact all developing or developed countries in the same way. The distributional impact will also vary within a country.[71]

S. Lonergan, The Role of Environmental Degradation in Population Displacement, 1998, p. 5; S. Castles, Environmental Change and Forced Migration: Making Sense of the Debate, 2002.

64 N. Stern, The Economics of Climate Change: The Stern Review, 2006, p. 129.

65 R. Black, Environmental Refugees: Myth or Reality?, 2001, p. 14.

66 Diverging view, see R. Black, Environmental Refugees: Myth or Reality?, 2001.

67 R. Zetter, The Role of Legal and Normative Frameworks for the Protection of Environmentally Displaced Persons, 2009, p. 398 *et seq.*

68 For an overview of various policy proposals, see S. C. McAnaney, Sinking Islands? Formulating a Realistic Solution to Climate Change Displacement, 2012, p. 1173 *et seq.*; similarly, K. M. Wyman, Responses to Climate Migration, 2013, p. 167 *et seq.*; S. F. Martin, Climate Change, Migration and Governance, 2010, p. 397 *et seq.*; E. Piguet, Climate Change and Formed Migration: How Can International Policy Respond to Climate Change Induced Displacement, 2008.

69 UNHCR, The Environment and Climate Change, 2015, p. 8.

70 IPCC, Climate Change 2001: Synthesis Report, p. 12.

71 IPCC, Climate Change 2001: Impacts, Adaptation, and Vulnerability, p. 957.

The assessment of the IPCC also highlights the particular vulnerabilities of Small Island States and indigenous peoples. Small Island States are particularly vulnerable, inter alia in view of their low elevation and small size, which puts them at the risk of inundation – partial or total – as a result of the rising sea level and more frequent and/or intense weather events such as cyclones. Indigenous peoples may be more exposed to the impact of climate change as a result of unique living conditions.[72] The concern at the particular adverse impact on indigenous peoples' rights and livelihoods has also been voiced by the Permanent Forum on Indigenous Issues: "Indigenous peoples have a vital interest in the development of international climate change policy and associated legal instruments, not least because they may be especially affected by climate change."[73] Indigenous peoples have sought to influence negotiations on international agreements related to climate change and, in particular, also in relation to policy and international law to reduce the impact of deforestation and environmental degradation.[74]

### 3 *Common but Differentiated Responsibility*

The concept of common but differentiated responsibility is closely related to climate justice, taking into account that the world's poorest and most vulnerable are the most affected by the impact of climate change, although they have contributed the least to this phenomenon.[75] The underlying assumption is based on the relationship between the level of industrialization of a country over time and climate change. The common but differentiated responsibility is a fundamental principle in any discussion addressing the challenges related to climate change, including climate displacement, although it does not necessarily meet agreement on all sides.

The principle first emerged at the Earth Summit in 1992 in Rio (i.e. the United Nations Conference on Environment and Development (UNCED)),[76] where States agreed that "[i]n view of the different contributions to global environmental degradation, States have common but differentiated responsibilities. The developed countries acknowledge the responsibility that they bear in the international pursuit of sustainable development in view of the pressures their societies place on the global environment and of the technologies and financial resources they command."[77] This principle was subsequently incorporated in the UNFCCC, the

72 *Ibid.*
73 E/C.19/2010/7.
74 OHCHR, Indigenous Peoples and the United Nations Human Rights System, Fact Sheet No. 9/Rev.2, 2013, p. 38.
75 UNDP, Human Development Report 2007/2008, p. 3; UNISDR, Global Assessment Report on Disaster Reduction, Risk and Poverty in Changing Climate, 2009, p. 10 *et subseq*.
76 For an assessment of UNCED, see O. Höll, Ökologisierung der internationalen Politik als Konfliktpotential: Die Konferenz über Umwelt und Entwicklung von Rio und die Folgen, 1993, p. 187 *et seq*.
77 A/CONF.151/26 (Vol. I), Principle 7.

first international legal instrument addressing climate change and its negative impacts. Article 3 of the UNFCCC stipulates that "[t]he Parties should protect the climate system for the benefit of present and future generations of humankind, on the basis of equity and in accordance with their common but differentiated responsibilities and respective capabilities. Accordingly, the developed country Parties should take the lead in combating climate change and the adverse effects thereof."[78] This is a recognition that climate change is a common concern relating to the notion in international law of *common heritage of mankind*, which was a concept introduced in the 1970s in relation to natural resources.[79] The differential treatment of developed and developing countries in the context of climate change is also a reflection of the principle of equity, while it does not absolve developing countries of their responsibility. The preamble of the UNFCCC emphasizes "that the global nature of climate change calls for the widest possible cooperation by all countries and their participation in an effective and appropriate international response, in accordance with their common but differentiated responsibilities and respective capabilities and their social and economic conditions."[80]

Controversies arise over the principle of common but differentiated responsibility when it comes to concrete pledges (i.e. climate targets during negotiations), including of the Paris Agreement. Developed countries hold the view that climate targets need to be set for all countries based on evolving global economic and national circumstances and not their historic contribution to climate change, taking into account the current and future contributions by emerging nations to carbon emissions.[81]

Climate displacement, which for most of its part will remain internal,[82] will in addition disproportionately affect developing countries, including SIDS.[83] These countries not only are among those that have least contributed to climate change, but also have the least capacity to address the effects of climate change, including climate displacement. An important implication thereof is the need for cooperation and international assistance without which many developing countries may not be in a position to effectively handle climate displacement, in particular when it comes to aspects such as relocation and the finding of durable solutions. The assistance

78 FCCC/INFORMAL/84, preambular para. 6.

79 A/RES/25/2749, operative para. 1.

80 FCCC/INFORMAL/84, preambular para. 6.

81 P. Pauw *et al.*, Different Perspectives on Differentiated Responsibilities: A State-of-the-Art Review of the Notion of Common but Differentiated Responsibilities in International Negotiations, 2014, p. 8 *et seq.*; for the discussion on the application of this principle to the Sustainable Development Goals framework, see C. Nobbe, Universality, Common but Differentiated Responsibilities and the Sustainable Development Goals, 2014, p. 6 *et seq.*

82 Government Office for Science, Foresight: Migration and Global Environmental Change, Final Project Report, 2011, p. 11; S. C. McAnaney, Sinking Islands? Formulating a Realistic Solution to Climate Change Displacement, 2012, p. 1178 *et subseq.*

83 Generally, *cf.* I. Schütz-Müller, Von der Entkolonisierung zu den gegenwärtigen Problemen der pazifischen Inselstaaten, 2009.

needed is not merely financial, but will include technical assistance and capacity building, to ensure that the burden is adequately shared by developed and developing countries, in line with their common but differentiated responsibility, to arrive at what is termed *climate justice* (i.e. solutions to climate change challenges that promote human rights, equity, labor rights and environmental justice globally and locally).[84]

## 4 *Displacement Patterns, Scenarios and Protection Implications*

While there are several classifications of disasters, the merit of the former Representative of the Secretary-General on the human rights of internally displaced persons, Walter Kälin, is that he associated the disaster types with concrete displacement patterns.[85] In a 2008 report,[86] he identified four climate change disaster types, which could coincide or overlap, and linked them to specific displacement patterns: (i) hydrometeorological disasters, which are projected to further increase in future and to lead to new and larger situations of displacement and often go hand in hand with destruction of property; (ii) general environmental degradation and slow onset disasters (i.e. the deterioration of conditions of life and economic opportunities in affected areas), which will incite voluntary population movements in an initial phase, and which could later transform into forced displacement and become permanent, as areas become less hospitable due to desertification or rising sea levels; (iii) sinking Small Island States, which will prompt internal relocation and migration abroad, including forced and permanent displacement; and (iv) climate change-induced armed conflict and violence triggered by a decrease in vital resources, attributable to climate change.[87]

These disaster categories can also be related to a range of displacement scenarios. Depending on whether displacement is of a temporary or permanent nature and whether displacement occurs within the territory or implies cross-border movements, with people remaining within the same region or moving to other continents, the following scenarios are identified: (i) temporary displacement as it occurs after hurricanes, floors or storms; (ii) permanent local displacement, if people are not able to return to their homes because or irreversible changes in their living environment;

84 The first Climate Justice Summit took place in 2000 in the Hague in parallel with COP 6; see *Guardian*, The first climate justice summit: A pie in the face for the global north, 2014; see also International Bar Association, Achieving Justice and Human Rights in an Era of Climate Disruption, 2014.

85 Renamed to Special Rapporteur on the human rights of internally displaced persons by Human Rights Council resolution 14/6 of June 17, 2010, A/HRC/RES/14/6, operative para. 11.

86 Representative of the Secretary-General on the human rights of internally displaced persons, Displacement Caused by the Effects of Climate Change: Who Will Be Affected and What Are the Gaps in the Normative Frameworks for their Protection?, 2008, p. 2.

87 *Ibid.*

(iii) permanent internal displacement; (iv) permanent regional displacement; and (v) permanent intercontinental displacement.[88]

In terms of protection, some of the disaster and displacement scenarios raise particular legal questions in connection with the applicable legal framework and States' responsibilities. Some movements prompted by climate change seem to fall within the existing protection framework, such as international refugee law or the protection of stateless persons, for example, the case of refugee flows provoked by climate change-induced armed conflict. International or regional refugee instruments and complementary forms of protection are applicable in this context, which also falls within the UNHCR's protection mandate.[89]

However, factual and legal difficulties arise when it comes to forced cross-border population movements, which do not result from internal violence or conflict. Individuals crossing international borders as a result of the erosion of their habitable areas do not fall within the existing refugee definition. Another lacuna exists in relation to populations that are forced to resettle, as could be the case for a number of low-lying SIDS that may be at risk of sinking.[90] Apart from the legal perspective relating inter alia to questions of statelessness, this also raises issues of preservation of cultural inheritance and cultural identity, which will not however be dealt with here.[91] The protection challenges arising from climate displacement will be discussed in more detail in Chapter 3 of this study.

88 S. Leckie, Climate-Related Disasters and Displacement: Homes for Homes, Lands for Lands, 2009, p. 4 *et subseq.*

89 UNHCR, Climate Change, Natural Disasters and Human Displacement: A UNHCR Perspective, 2009, p. 2 *et subseq.*

90 CARE *et al.*, In Search of Shelter: Mapping the Effects of Climate Change on Human Migration and Displacement, 2009, p. 19; for an in-depth discussion, see, e.g., S. Park, Climate Change and the Risk of Statelessness: The Situation of Low-Lying Island States Susin Park, 2011; J. McAdam, "Disappearing States," Statelessness and the Boundaries of International Law, 2012, p. 105 *et seq.*

91 *Cf.* I. Schütz-Müller, Zur Funktion von Intellektuellen in pazifischen Inselstaaten, 2000.

# 2

# Connecting the Dots

Against this background, the ensuing chapter will look at how the dots connected, that is, how the climate change and displacement nexus became progressively recognized and, in parallel, how the impact of climate change on human rights was acknowledged within the UN human rights machinery and subsequently in the climate change framework itself.

## I HUMAN MOBILITY AND THE CLIMATE CHANGE FRAMEWORK

### 1 *The Intergovernmental Panel on Climate Change*

The IPCC, established in 1988 by the World Meteorological Organization (WMO) and the United Nations Environment Programme (UNEP)[1] to assess the information relevant to the scientific basis of the risk of human-induced climate change, its potential impact and possible response strategies, ascertained, in its first assessment report in 1990, that the gravest impact of climate change may be on human migration.[2] It has since issued several assessment reports (1995, 2001, 2007 and 2014), special reports, technical papers and methodology reports, which are standard references in climate change research and inform the decisions of policy-makers and experts worldwide. It is important to note however that these assessments are not apolitical, as they are based on consensus – which does not bode well with the principles of scientific research – and are moreover subject to review and approval by the intergovernmental body.[3]

Warning initially, in 1990, that millions would be displaced as a result of the effects of climate change, the IPCC subsequently changed its perspective on climate change-induced migration, adopting a more nuanced and prudent position on this issue in view of the difficulties to make incontrovertible predictions on migrant decisions, including those related to climate change effects. The IPCC's focus

1 A/RES/43/53 of December 6, 1988, para. 5.
2 IPCC, First Assessment Report, 1990, p. 103.
3 D. Kniveton *et al.*, Climate Change and Migration: Improving Methodologies to Estimate Flows, 2008, p. 16 *et subseq.*

shifted from migration to adaptive capacities and vulnerabilities of populations, mindful of the multifaceted ways in which people may adapt to climate change and of the difficulties to separate climate change as a root cause of displacement from coinciding, complementary or alternative causes, such as conflict.[4] In its 2007 report, the IPCC therefore addressed the potential of population migration only as a result of two particular manifestations of climate change (i.e. in relation to droughts and cyclones) and not as an issue in itself.[5] The report concludes that drought and the consequent loss of livelihoods is a major trigger for population movements, particularly from rural to urban areas.[6] However, while the IPCC notes that "[c]limate change and sea-level rise affect coastal settlements and infrastructure,"[7] it does not infer that this would directly result in population movements. These are rather considered as consequences of the related effects on the health of those concerned.[8]

**Initial Mandate of the IPCC**

The General Assembly requested the Secretary-General of the World Meteorological Organization and the Executive Director of the Organization and the United Nations Environment Programme, utilizing the Intergovernmental Panel on Climate Change, immediately to initiate action leading – as soon as possible – to a comprehensive review and recommendations with respect to:

- The state of knowledge of the science of climate and climatic change;
- Programmes and studies on the social and economic impact of climate change, including global warming;
- Possible response strategies to delay, limit or mitigate the impact of adverse climate change;
- The identification and possible strengthening of relevant existing international legal instruments having a bearing on climate change;
- Elements for inclusion in a possible future international convention on climate.

*Source: United Nations General Assembly resolution 43/53 of December 6, 1988*

It was only in its most recent assessment report, in 2014, that the IPCC for the first time examined in detail the aspects of human mobility as a result of climate change. More than twenty years after its initial affirmation of a link between climate change

4 IPCC, Climate Change 2001: Synthesis Report, 2001, p. 3.

5 IPCC, Climate Change 2007: Impacts, Adaptation and Vulnerability, Summary for Policymakers, p. 18.

6 IPCC, Climate Change 2007: Impacts, Adaptation and Vulnerability, 2007, p. 399.

7 *Ibid.*, p. 333.

8 IPCC, Climate Change 2007: Impacts, Adaptation and Vulnerability, Summary for Policymakers, 2007, p. 18.

and human mobility, the IPCC's report concludes that "[c]limate change is projected to increase displacement of people."[9] The IPCC also affirms that "[c]hanges in migration patterns can be responses to both extreme weather events and longer-term climate variability and change, and migration can also be an effective adaptation strategy."[10] It is finally with this report, after nuancing its position since 1990 and avoiding to make any assertions, that the IPCC consolidates its position on the question of human mobility relating to climate change. From the scientific angle, the nexus between climate change and population movements has now been established, which is essential for the political level to follow suit.

**Climate Framework Key Dates**

| | |
|---|---|
| 2016 | At COP 21 in Paris, on December 12, 2015, Parties to the UNFCCC adopt the landmark Paris Agreement. |
| 2015 | Adoption of the Paris Agreement on December 12, 2015. |
| 2014 | At COP 20 in Lima, Parties adopt the Lima Call for Action, which elaborates key elements of the forthcoming agreement in Paris. |
| 2010 | Cancún Agreements adopted. |
| 2005 | Entry into force of the Kyoto Protocol. |
| 1997 | The Kyoto Protocol is adopted at COP 3 in December. It seeks to reduce greenhouse gas concentrations in the atmosphere to "a level that would prevent dangerous anthropogenic interference with the climate system" (Art. 2). |
| 1995 | The Conference of the Parties (COP) becomes the UNFCCC's governing body, and the first Conference of the Parties (COP 1) takes place in Berlin. |
| 1994 | The UNFCC enters into force. |
| 1992 | At the Earth Summit in Rio de Janeiro, the UNFCCC is opened for signature. |
| 1990 | First IPCC assessment report. UN General Assembly negotiations on a framework convention begin. |
| 1988 | Establishment of the IPCC. |
| 1980–1990 | A number of intergovernmental conferences focus on climate change. |
| 1979 | First World Climate Conference (WCC) calls on governments to address climate change. |

[9] IPCC, Climate Change 2014: Synthesis Report, 2014, p. 73.

[10] IPCC, Climate Change 2014: Impacts, Adaptation and Vulnerability, Summary for Policymakers, 2014, p. 20.

## 2 *United Nations Framework Convention on Climate Change*

A different framework is the UNFCCC, which was adopted in 1992 as the first international treaty on climate change. The UNFCCC provides a framework for international cooperation to combat climate change by stabilizing greenhouse gas concentrations in the atmosphere at a level that would prevent dangerous anthropogenic interference with the climate system (i.e. ultimately to limit average global temperature increases and the resulting climate change), as well as to cope with its impacts.[11] Therefore, by contrast to the IPCC, the UNFCCC framework is concerned with anthropogenic changes to the climate only.

The two frameworks, while operating independently from each other from an institutional and legal perspective, are interrelated to the extent that IPCC assessments play a fundamental role to inform policy development and adoption at the level of the UNFCCC Conferences of the Parties, which are normally held on an annual basis,[12] through their authoritative expert advice.

The UNFCCC's main focus lies on mitigation of climate change by reducing greenhouse gas emissions and enhancing sinks.[13] In this regard, by adopting the Kyoto Protocol in 1998, States committed themselves to adopting national policies and taking corresponding measures on the mitigation of climate change and to reducing their overall greenhouse gas emissions.[14] The UNFCCC also recognizes the need for adaptation to the impacts of climate change, but places the emphasis on the natural adaption of ecosystems to climate change, with the overall objective to ensure that food production is not threatened and to enable economic development to proceed in a sustainable manner.[15] This perspective does not take into account human mobility as a possible form of adaptation.

The primary focus on mitigation is a logical consequence of the nature of the convention and the objectives of the UNFCCC, but has been somewhat outdated by reality and scientific findings. The IPCC in its Working Group I Fourth Assessment Report concluded that climate change cannot be reversed and that therefore "[i]rrespective of the scale of mitigation measures, adaptation measures are necessary" to address the impacts of climate change.[16]

11 For further background information on the UNFCCC, see http://unfccc.int/files/essential_background/background_publications_htmlpdf/application/pdf/conveng.pdf (accessed May 4, 2019).

12 COP meetings have been held every year since 1995 with the exception of 2001, when COP meetings took place in Bonn and in Marrakech within the same year. See http://unfccc.int/meetings/items/6237.php?filtbody=53 (accessed May 4, 2019).

13 The IPCC defines mitigation as "[a]n anthropogenic intervention to reduce the sources or enhance the sinks of greenhouse gases." See IPCC, Climate Change 2001: Synthesis Report, 2001, p. 379.

14 Kyoto Protocol to the United Nations Framework Convention on Climate Change, adopted on December 11, 1997.

15 See Kyoto Protocol to the United Nations Framework Convention on Climate Change, Art. 2.

16 IPCC, Climate Change 2007: Mitigation, Summary for Policymakers, 2007, p. 21.

In spite of any explicit reference in the UNFCCC, the Kyoto Protocol or any decisions made by the COP to the UNFCCC that would consider adaptation and mitigation as potential alternatives,[17] the UNFCCC framework in 2010 eventually recognized the need for both, mitigation and adaptation. For the first time, it explicitly considered human mobility implications, acknowledging that adaptation[18] must be addressed with the same level of priority as mitigation. The Climate Change Conference in Cancún, Mexico, in 2010 (COP 16) adopted the Cancún Adaptation Framework (CAF) as part of the Cancún Agreements and explicitly recognized migration, displacement and planned relocation as forms of adaptation to climate change.[19] The Agreements moreover for the first time established a link between the climate change and human rights frameworks by making reference, in its preamble, to Human Rights Council resolution 10/4 on human rights and climate change, "which recognizes that the adverse effects of climate change have a range of direct and indirect implications for the effective enjoyment of human rights and that the effects of climate change will be felt most acutely by those segments of the population that are already vulnerable owing to geography, gender, age, indigenous or minority status, or disability."[20] The 2012 Doha decision is also to be mentioned, attesting to the gradual entry of the human mobility question into the climate change framework, as it encourages "further work to advance the understanding of and expertise on loss and damage, which includes [. . .] enhancing the understanding of [. . .] how impacts of climate change are affecting patterns of migration, displacement and human mobility."[21]

It is also worth mentioning that the progressive recognition of displacement as a form of adaption, and the call in the Cancún Agreements to take measures "to enhance understanding, coordination and cooperation with regard to climate change induced displacement, migration and planned relocation,"[22] gave an important impetus for the work done by Displacement Solutions on climate displacement that culminated in the elaboration of the Peninsula Principles on Climate Displacement within States in 2013.

[17] IPCC, Inter-relationships between adaptation and mitigation, in: Climate Change 2007: Impacts, Adaptation and Vulnerability.

[18] The IPCC defines adaptation as "[a]djustment in natural or human systems in response to actual or expected climatic stimuli or their effects, which moderates harm or exploits beneficial opportunities." See IPCC, Climate Change 2001: Synthesis Report, 2001, p. 365.

[19] FCCC/CP/2010/7/Add.1, March 15, 2011, Decision 1/CP.16, The Cancun Agreements: Outcome of the work of the Ad Hoc Working Group on Long-term Cooperative Action under the Convention, para. 2(b).

[20] *Ibid.*, preamble.

[21] FCCC/CP/2012/8/Add.1, February 28, 2013, 3/CP.18, Approaches to address loss and damage associated with climate change impacts in developing countries that are particularly vulnerable to the adverse effects of climate change to enhance adaptive capacity, para. 7(a)(vi).

[22] FCCC/CP/2010/7/Add.1, March 15, 2011, Decision 1/CP.16, The Cancun Agreements: Outcome of the work of the Ad Hoc Working Group on Long-term Cooperative Action under the Convention, para. 2(b).

## 3 *The Paris Agreement*

Building upon the UNFCCC, the Paris Agreement (COP 21) adopted on December 12, 2015 – even though it failed on very important fronts – was a breakthrough in terms of the international recognition not only of the nexus between climate change and human mobility, but also of the need to address climate displacement. The Agreement entered into force on November 4, 2016.[23]

In its preamble, the Agreement contains an important reference to migrants acknowledging

> that climate change is a common concern of humankind, Parties should, when taking action to address climate change, respect, promote and consider their respective obligations on human rights, the right to health, the rights of indigenous peoples, local communities, migrants, children, persons with disabilities and people in vulnerable situations and the right to development, as well as gender equality, empowerment of women and intergenerational equity.[24]

Another key element with regard to the issue of displacement and population movements is the establishment of a Climate Displacement Task Force "to complement, draw upon the work of and involve, as appropriate, existing bodies and expert groups under the Convention as well as relevant organizations and expert bodies outside the Convention, to develop recommendations for integrated approaches to avert, minimize and address displacement related to the adverse impacts of climate change."[25] While this has been an important achievement, the Task Force is not however part of the Paris Agreement, which, as an international treaty, is of a legally binding nature, but only of the decision adopting the Paris outcome, thereby reflecting a certain degree of political commitment.

As a matter of fact, the Task Force was conceived since no agreement could be reached on the Climate Displacement Facility that was envisaged by the initial draft negotiation text. The mandate envisaged for the Climate Change Displacement Coordination Facility was "to help coordinate efforts to address climate change induced displacement, migration and planned relocation."[26] This objective of the facility would have included support for emergency relief, but also more concretely assistance for an organized migration and planned relocation. Another aspect was compensation measures for climate displaced persons.[27] The agreement on its establishment per se nevertheless constitutes an important recognition of the nexus between climate change and displacement and the challenges associated with it.

23 For the status of ratification, see http://unfccc.int/paris_agreement/items/9444.php (accessed May 10, 2019).

24 FCCC/CP/2015/10/Add.1, Annex, Paris Agreement.

25 FCCC/CP/2015/10/Add.1, Decision 1/CP.21, Adoption of the Paris Agreement, operative para. 49.

26 FCCC/ADP/2015/L.6, December 5, 2015, Art. 5. For further details, see J. Wentz, M. Burger, Designing a Climate Change Displacement Coordination Facility: Key Issues for COP 21, 2015.

27 FCCC/ADP/2015/1, February 25, 2015, para. 32 and Option III, para. 33.

The Climate Displacement Task Force, which was operationalized in March 2017 and held the first of two in-person meetings in May 2017, offered a key opportunity within the UN framework to collect information and formulate concrete recommendations on how to address the displacement challenges resulting from climate change. The Task Force faced the difficult task to overcome the political controversies inherent to its work, notably on the causal relations between climate change and displacement. It was also tasked at COP 23 (November 2017), by its decision 5/CP.23, to take into consideration both cross-border and internal displacement. The Task Force submitted its recommendations in September 2018 for discussion at a meeting of the WIM, the formal mechanism at the UNFCCC for addressing the loss and damage caused by climate change, and while its mandate was technically supposed to end after it had to present its recommendations at COP 24 in Katowice in December 2018, it was ultimately extended sine die to continue its work. The recommendations, which were endorsed at COP 24 as an annex to the WIM's final text, invite the Parties "to consider formulating laws, policies and strategies, as appropriate, that reflect the importance of integrated approaches to avert, minimize and address displacement related to the adverse impacts of climate change and in the broader context of human mobility, taking into consideration their respective human rights obligations and, as appropriate, other relevant international standards and legal considerations."[28] From a climate displacement perspective this was the most important outcome of Katowice. There has also been a strong push toward integrating a human rights-based approach into the implementation guidelines for the Paris Agreement, notably with the – even though brief – participation of the UN High Commissioner for Human Rights at the conference.[29] Thirty-four UN Human Rights Special Procedures had also issued a joint letter calling for governments to integrate human rights considerations in the implementation guidelines.[30] This has not however allowed to secure mainstreaming of human rights, as the sole reference to human rights was removed from the text in the final stage of the negotiations.

## 4 *Climate Change and the Human Mobility Framework*

The international institutional human mobility framework currently consists of two lead organizations (i.e. the UNHCR and the IOM), which both emerged after World War II to deal with a massive population of displaced persons and refugees. Climate displaced persons, depending on the concrete circumstance, may squarely

28 FCCC/SB/2018/1, Report of the Executive Committee of the Warsaw International Mechanism for Loss and Damage associated with Climate Change Impacts.

29 Michelle Bachelet, UN High Commissioner for Human Rights, statement at the 24th Conference of the Parties to the UN Framework Convention on Climate Change, Global Commission on Adaptation, Accelerating Action and Support for Adaptation, Katowice, December 12, 2018.

30 PR/OHCHR/6December2018, Joint statement of the United Nations Special Procedures Mandate Holders on the occasion of the 24th Conference of the Parties to the UNFCCC.

fall within the mandate of either organization or neither of them. Both the UNHCR and the IOM recognized the challenges raised by climate displacement, overall and for their organizations, as well as the potential protection gaps at the international and national levels.

### a United Nations High Commissioner for Refugees

The Office of the UNHCR was established in 1949 by General Assembly resolution 428 (V). On December 14, 1950, the General Assembly proceeded with the adoption of the Statute of the Office of the UNHCR as an Annex to resolution 428 (V). Paragraph 2 of the Statute provides that "[t]he work of the High Commissioner shall be of an entirely non-political character; it shall be humanitarian and social and shall relate, as a rule, to groups and categories of refugees."[31]

The initial mandate of the UNHCR encompassed the protection of refugees and the promotion of durable solutions to address their situation. Coordinating resettlement efforts therefore became the UNHCR's primary focus in its early years. While not laid down in its initial mandate, as major refugee crises started occurring in developing countries, the coordination of material assistance for refugees, returnees and displaced persons, which had hitherto been seen as the primary responsibility of the governments concerned, became gradually one of the principle functions of the UNHCR. When in 1967, with the adoption of the Protocol relating to the Status of Refugees, the restrictions of the UNHCR's mandate, which had until then only extended to persons displaced by World War II or related to the immediate postwar period, were removed, the UNHCR started its worldwide engagement. The General Assembly further decided in 2003 to remove the temporal limitation on the continuation of the UNHCR, granting it a permanent mandate "until the refugee problem is solved."[32] Starting out as a protective agency with a focus on representing refugees legally and diplomatically, whereby the final determination of the refugee status remained the primary responsibility of the State, its functions over time increasingly entailed the provision of emergency relief and humanitarian assistance as well as addressing the root causes of the refugee problems.[33]

Whereas internally displaced persons were not originally included in the mandate of the UNHCR, as their protection was part of States' area of competence and responsibility, many of which were concerned over external interference into domestic affairs, the UNHCR had been mandated since 1976 by the General Assembly – first on a case-by-case basis – to assume responsibility for internally displaced persons in certain situations. At present, General Assembly Resolution 53/

[31] UNHCR, Statute of the Office of the United Nations High Commissioner for Refugees, Annex to General Assembly resolution 428 (V) of December 14, 1950.

[32] A/RES/58/153 of December 22, 2003, para. 9.

[33] D. P. Forsyth, Human Rights in International Relations, 2012, p. 99 *et subseq.*

125 of December 9, 1998, is seen as the legislative basis for the UNHCR's involvement in internally displaced persons' protection. In paragraph 16, the Assembly

> reaffirms its support for the role of the Office of the High Commissioner in providing humanitarian assistance and protection to internally displaced persons, on the basis of specific requests from the Secretary-General or the competent organs of the United Nations and with the consent of the State concerned, taking into account the complementarities of the mandates and expertise of other relevant organizations, and emphasizes that activities on behalf of internally displaced persons must not undermine the institution of asylum.[34]

Recognizing the complexities of the dynamics of displacement in general and the difficulty to separate the underlying causes of displacement in particular, the UNHCR, as early as 1993, in its State of the World's Refugees Report, addressed the question of climate displacement. In its approach, rather than emphasizing the causal relationship between climate change and displacement, it merely added environmental roots to the possible and observed displacement causes, without entering however into the debate whether displacement was attributed exclusively to climate change. The report stated:

> Millions of people have been forced to leave their homes because the land on which they live has become uninhabitable or is no longer able to support them. In some cases the cause is a natural disaster; in others, the catastrophe is caused by humans. The disruption to the habitat may be sudden, as at Chernobyl or Mount Pinatubo, or as gradual as the spread of a desert or the retreat of a forest.[35]

As a result, depending on the particular circumstances of the case, certain categories of internal or cross-border climate displaced persons may fully or to some extent fall within the scope of the UNHCR's mandate.

### b International Organization for Migration

Climate migrants, in accordance with its diction, may fall within the mandate of the IOM, which was established as a result of the Migration Conference in Brussels held in December 1951, initially to facilitate the resettlement of the so-called surplus populations from Europe to other countries overseas, whereby these countries retained control of standards of admission and of the number of immigrants to be admitted.[36] Today, it is the lead intergovernmental organization on migration issues, and since the early 1990s it has been engaged on the issue of climate migration and

34 A/RES/53/125 of December 9, 1998, para. 16.

35 UNHCR, The State of the World's Refugees 1993, chapter 1.

36 Intergovernmental Committee for European Migration (ICEM), Resolution to Establish a Provisional Intergovernmental Committee for the Movement of Migrants from Europe, 1951; M. Ducasse-Rogier, The International Organization for Migration, 1951–2001.

displacement and has been mandated by its Member States to more closely consider this issue.[37]

Recognizing the need for a comprehensive approach to human mobility, including both migrants and refugees, and for enhanced cooperation in this regard at the global level, heads of state gathered for a high-level plenary meeting on addressing large movements of refugees and migrants at the 70th session of the General Assembly on September 19, 2016. On this occasion, in line with General Assembly resolution 70/296 of July 25, 2016,[38] the IOM was brought "into a closer legal and working relationship with the United Nations as a related organization."[39]

The New York Declaration for Refugees and Migrants adopted further to the high-level plenary meeting, initiated an intergovernmental process toward the elaboration of a global compact for safe, orderly and regular migration, which was adopted at an intergovernmental conference on international migration in December 2018.[40] The Declaration recognized among the many causes of human mobility that people move "in response to the adverse effects of climate change, natural disasters (some of which may be linked to climate change), or other environmental factors"[41] and contained numerous references to human rights. The coordination of the work of the United Nations to implement the New York Declaration was entrusted to Louise Arbour, who was appointed Special Representative of the Secretary-General for International Migration on March 9, 2017.[42] As she had previously been the UN High Commissioner for Human Rights, hopes were high that her nomination would facilitate the mainstreaming of a human rights-based approach into the future global compact.

The UN Global Compact recognizes "the adverse effects of climate change" as a distinct driver of forced migration (Objective 2, paragraphs 19.h–19.l). This is a significant change from the zero draft of January 2018, which did not directly refer to climate change, but was rather listing related causes for forced migration, such as "sudden on-set natural disasters, slow-onset environmental degradation, emergency situations, and other life-endangering circumstances." While it thus recognizes the challenges linked to the environment–migration nexus, it does not however adequately respond to climate displacement, as it lumps together various migrants in vulnerable situations, which includes climate-induced displaced persons. It nevertheless acknowledges that many movements occur as result of a combination of reasons and that climate change and environmental factors interact with other migration drivers. The Global Compact is mainly concerned with cross-border migration, not with internal displacement, which is evidenced by

37 See O. Brown, Migration and Climate Change, 2008.
38 A/RES/70/296 of July 25, 2016.
39 A/RES/71/1 of September 19, 2016, para. 49.
40 A/RES/71/1 of September 19, 2016.
41 *Ibid.*, para. 1.
42 SG/A/1712 of March 9, 2017.

Objective 5, which concedes that adaptation *in situ* may not be possible in certain circumstances and that planned relocation and visa options need to be part of migration management tools. The need to prioritize climate change mitigation and adaptation measures in countries of origin is also seen through the angle of minimizing migration drivers rather than the necessity to prevent or address internal displacement.[43]

Similarly, the Global Compact on Refugees, which was also adopted in December 2018, enhances the protection framework in the context of climate change and disasters by recognizing that "environmental degradation and natural disasters increasingly interact with the drivers of refugee movements."

### 5 *Intermediate Remarks*

It took an incredible twenty years for the initial affirmation of the IPCC that climate change would lead to massive displacement to be finally recognized and firmly anchored in the UNFCCC framework. The acknowledgment of the nexus between climate change and displacement on the one hand and the recognition that this would constitute an adaptation strategy for affected populations on the other hand served as important catalysts for normative and policy initiatives to emerge. This allowed moving beyond the level of policy development to the level of implementation, action-oriented guidelines and the development of appropriate institutional arrangements.

## II CLIMATE CHANGE AND HUMAN RIGHTS: THE PATH OF CONVERGENCE

### 1 *Pioneering the Nexus: Ksentini Reports*

When climate change started increasingly to gain visibility on the international agenda, for instance with the creation of the IPCC in 1998, the human rights machinery also began being seized of the matter. It was the former Sub-Commission on Prevention of Discrimination and Protection of Minorities (i.e. the United Nations' expert body of the former Commission on Human Rights)[44] – which functioned as the actual think tank of the UN human rights machinery – that, as the first human rights body, decided in August 1989 by its decision 1989/108 to task one of its members, Fatma Zohra Ksentini, to study the "problem of the environment and its

43 A/CONF.231/3.

44 The UN Sub-Commission on the Promotion and Protection of Human Rights, the main subsidiary body of the former Commission on Human Rights, was established in 1947 with twelve members. Originally the Sub-Commission on the Prevention of Discrimination and Protection of Minorities, it was renamed in 1999. The Sub-Commission on the Promotion and Protection of Human Rights was replaced by the Advisory Committee. For further information on the Sub-Commission, see www.ohchr.org/EN/HRBodies/SC/Pages/SubCommission.aspx (accessed March 26, 2019).

relation to human rights."[45] This decision was of major significance because it meant a change of paradigm. Until then, for many years, environmental problems were almost exclusively looked at from a protection-from-pollution perspective and the responsibilities arising in this regard. Recognizing, however, that there might be a linkage between human rights and the environment was a new understanding, accepting reciprocal effects and interactions between environmental questions and human rights, therefore tackling the issue from an inclusive angle by adding the human dimension.

This significant decision materialized not least on account of proactive, well resourced and also procedurally versed civil society actors.[46] The NGOs involved did not merely call for this decision,[47] but disposing over substantive expertise, managed by virtue of their actions and submissions to make an important contribution to the various processes, thus creating momentum, which ultimately led to the formulation and adoption of the decision. In fact, some NGOs provided enabling assistance as well as endorsement with an overall commitment and their particular support for the study prepared by Special Rapporteur Ksentini. They also assisted in organizing international meetings, seminars and consultations in support of the Special Rapporteur's work, the deliberations and outcomes of which fed into the Rapporteur's report, and also meaningfully contributed to the report itself.[48] This is of particular importance as it suggests that NGOs, along with the expert members of the Sub-Commission, States, UN organs and specialized agencies, had a persuasive role in this context being the voices of the peoples, articulating their concerns and proposed solutions and conveying them to this UN body. NGOs through their various contributions helped analyze the issue and provided the international community and the United Nations with a diagnostic apt to engender, through evidence-based information and expert analysis, awareness, persuasion and ultimately strategic and anticipatory decisions on measures in law and policy potentially facing challenges in the future in connection with environmental problems and their impact on human rights.

45 E/CN.4/Sub.2/1994/9 of July 6, 1994, para. 8 *et subseq*.

46 Such as the Friends of the Earth, the Sierra Club Legal Defense Fund, the Association of Humanitarian Lawyers – see E/CN.4/Sub.2/1994/9 of July 6, 1994, para. 8; for an overview of the environmental rights, see Earthjustice, Environmental Rights Report, Human Rights and the Environment, Materials for the 61st session of the United Nations Commission on Human Rights, Geneva, March 14 – April 22, 2005.

47 See, e.g., NGO statement in E/CN.4/Sub.2/1989/SR.18 of August 18, 1989, 1. Karen Parker (Friends of the Earth), speaking on behalf of her own organization, the Sierra Club Legal Defence Fund and the Association of Humanitarian Lawyers, drew attention to the relationship between human rights and the environment. As a result of this intervention, the Sub-Commission adopted decision 1989/108 asking Ksentini to prepare a note.

48 E/CN.4/Sub.2/1994/9, para. 17 *et subseq*. Therein the Special Rapporteur Ksentini thanks explicitly the Sierra Club Legal Defense Fund for all its support, including a seminar in Geneva on May 15–19, 1994, for the purpose of drawing up the guiding principles annexed to this final report on the human rights and the environment; see also Earthjustice, Environmental Rights Report, Human Rights and the Environment, Materials for the 61st session of the United Nations Commission on Human Rights, Geneva, March 14 – April 22, 2005.

Thereupon, the Sub-Commission requested Ksentini, one of its expert members, to prepare a note on methods to prepare a study on the relationship between human rights and the environment.[49] The approach taken is of interest, as the then Sub-Commission, now Advisory Committee of the United Nations Human Rights Council, decided in a first step merely to examine the methodology for the preparation of a more in-depth study.[50] At the following session, the Commission on Human Rights welcomed the Sub-Commission's decision. In its resolution 1990/41 of March 6, 1990, "Human rights and the environment," it furthermore underscored the link between preservation of the environment and the promotion of human rights.[51]

The Special Rapporteur's note entitled "Human Rights and Scientific and Technological Developments: Proposals for a study of the problem of the environment and its relation to human rights," dated August 3, 1990, contained some preliminary analytic and methodological considerations for the preparation of a study. In addition, it discussed conceptual aspects and included an outline of the emergence of a right to the environment as well as a presentation of the main topics on which to focus the study, before closing with a number of recommendations.[52]

While displacement issues were not explicitly mentioned, paragraph 7 of the note takes account of its potential incompleteness by adding an exculpation clause that reads as follows: "In view of the constraints that affect any brief note on a broad issue entailing complex factors and with a changing legal basis, it is impossible for the contents of this document to be seen as other than a set of working hypotheses that require deeper consideration, shaping of ideas and fine-tuning of the proposed approach."[53]

While not anticipating at this stage any (massive) climate displacement scenarios, the note mentions nevertheless the likelihood of natural habitats, natural resources and human habitat and settlements (housing, town-planning, demography etc.) to be negatively affected, and that particular emphasis could be laid on the essential factors involved, namely aspects linked to human habitats and settlements, and the rights of the individual.[54]

It is also clear from this note that the study mainly aimed to highlight the human rights impact of environmental problems. An emphasis was put on those rights that seemed prospectively to conflict with the requirements of environmental protection.[55]

49 E/CN.4/Sub.2/1989/58, at para. 71.

50 E/CN.4/Sub.2/1994/9, para. 9 *et subseq.*

51 At its 42nd session containing a number of proposals for a study of the environment and its relation to human rights, see E/CN.4/Sub.2/1990/12 in conjunction with the final report, E/CN.4/Sub.2/1994/9 at para. 12.

52 E/CN.4/Sub.2/1990/12.

53 *Ibid.*, para. 7.

54 *Ibid.*, paras. 11 and 26.

55 *Ibid.*, para. 27.

After presentation of her note to the Sub-Commission in August 1990, and upon its careful examination, the United Nations body appointed Ksentini as the first ever Special Rapporteur on human rights and the environment. By virtue of Sub-Commission resolution 1990/7 dated August 30, 1990, the Rapporteur was formally tasked to prepare and submit a preliminary report on the issue at its next session. This course of action was endorsed by the Commission on Human Rights through its resolution 1991/44 dated March 5, 1991.[56]

At this stage, the question of displacement emerged but marginally in relation to indigenous peoples through Sub-Commission resolution 1990/27, which invited the Special Rapporteur also to reflect on the specific relationship between fragile habitats and indigenous peoples, especially with respect to sustainability.

In line with the agreed methodology, the Special Rapporteur submitted a preliminary report,[57] which was considered and endorsed by the Sub-Commission requesting the Rapporteur to prepare a progress report examining inter alia relevant national and international provisions.[58] Again, the Commission on Human Rights[59] and the Economic and Social Council,[60] as the respective parent bodies, thereon endorsed this request.

The Sub-Commission examined the progress report,[61] and the review and analysis of national and international provisions therein, as well as relevant decisions and comments of human rights bodies relating to human rights and the environment, with a view to detect and identify environmental rights as human rights. In the same vein, Ksentini had stated in her report that the relationship between the environment and human rights implicated other rights, in particular the rights to life, culture and health.[62]

In spite of explicit references to climate displacement, NGOs brought forward the issue of environmental refugees, stressing that this was a human rights abuse problem and outlining that according to estimates, by the end of the century (i.e. the twentieth century), the number of environmental refugees would exceed 100 million worldwide. Accordingly, the most vulnerable areas were in Africa and Asia. There was also an affirmation that "[t]he environmental refugee problem has now reached crisis proportions and is only likely to deepen because these victims are not protected by international law and are ineligible to receive humanitarian assistance under traditional refugee programs."[63]

56 See Commission on Human Rights resolution 1991/44 of March 5, 1991, which was endorsed by ECOSOC decision 1991/244 of May 31, 1991.
57 E/CN.4/Sub.2/1991/8.
58 E/CN.4/Sub.2/1991/24, Sub-Commission resolution 1991/24 of August 29, 1991.
59 Commission on Human Rights decision 1992/110 of February 28, 1992.
60 ECOSOC decision 1992/252 of July 29, 1992.
61 E/CN.4/Sub.2/1992/7 and Add.1.
62 In addition, the reports E/CN.4/Sub.2/1992/7 and Add.1 addressed the United Nations Conference on Environment and Development (UNCED) held in Brazil, June 3–14, 1992, and noted in an addendum relevant principles in the Rio Declaration on Environment and Development.
63 E/CN.4/Sub.2/1992/SR.31, para. 9.

By virtue of resolution 1992/31 dated August 27, 1992, the Sub-Commission in its diction "took note with appreciation" of the progress report and requested the Special Rapporteur to continue her study and submit a second progress report containing additional information and an analysis of decisions and views of international human rights and other related organs as well as information on and analysis of national laws and practice.[64]

In this second progress report,[65] Ksentini reviewed developments in regard to the recognition and implementation of environmental rights as human rights on the basis of the standards and practices developed at the national, regional and universal levels. The second progress report seemed to have come in a different shape as it stipulated a kind of content proviso by stating that it would not revert to analyses made in earlier reports.[66] As a matter of fact, in the context of evolutions at the regional level, the report mentions the OECD, which submitted guidelines from its Development Assistance Committee, of which Guidelines No. 3: Guidelines for Aid Agencies on Involuntary Displacement and Resettlement in Development Projects "[call] on project designers and implementors to ensure that the population displaced by a project receives benefits from the changes and that it is re-established on a sound productive basis."[67] With respect to resettlement planning, it should take into account basic policy consideration; inter alia, "[p]articipation by environmental agencies and community participation in planning and implementing resettlement is essential and should include women."[68]

For the first time, moreover, climate displacement and the aspects relating to the mandate of the Office of the UNHCR were raised. The report outlined that

> it is becoming increasingly evident that environmental degradation can contribute, directly or indirectly, to population displacements, including refugee flows, and that the international community should take steps to address this phenomenon. As the High Commissioner stated at the Conference on Environment and Development in June 1992, "research should be undertaken to investigate the interaction between environmental and socio-economic factors, and the possible implications of this relationship for migration."[69]

In the final section of her report, Ksentini recommended that the various human rights bodies should examine, in their respective fields of concern (i.e. children, women, economic, social and cultural rights et al.), the environmental dimension of those human rights under their responsibility. She continued to suggest that in

64 The Commission on Human Rights endorsed this request in its decision 1993/144 of March 10, 1993, E/CN./Sub.2/1993/144.

65 E/CN.4/Sub.2/1993/7.

66 *Ibid.*, para. 9.

67 OECD, Development Assistance Committee Guidelines on Aid and Environment, No. 3, Guidelines for Aid Agencies on Involuntary Displacement and Resettlement in Development Projects, 1992.

68 E/CN.4/Sub.2/1993/7, para. 56.

69 *Ibid.*, para. 104.

particular the bodies, committees, working groups and special rapporteurs concerned should regularly examine the topic of environmental refugees and displaced persons.[70]

The final report of the Sub-Commission on Prevention of Discrimination and Protection of Minorities[71] noted that "[f]or many years environmental problems were almost exclusively considered from the standpoint of the pollution in one part of the world [...]. Acknowledgement of the link between the environment and human rights was fostered by an awareness of the global, complex, serious and multidimensional nature of environmental problems."[72]

Ksentini's final report provided a new state of knowledge, going beyond a limited framework and narrow vision that so far circumscribed the analysis of environmental problems, and suggested a more comprehensive approach, involving global economic, social and cultural aspects to which it added the human dimension.[73]

This human dimension, besides refugees and internally displaced persons,[74] captures for instance the situation of children, in view of their particular vulnerability, in accordance with Article 22 of the Convention on the Rights of the Child regarding refugee children.[75] In the same vein, situations of climate or environmental displacement leading to the removal of indigenous peoples from their lands and/or their displacement or preempting them from using their lands are of pertinent relevance. Consequently, the human rights concerns resulting from displacement and/or degradation of traditional indigenous lands is raised in this context.[76]

A fundamental facet requiring a notional discussion and anticipatory action is, as Ksentini reiterated in her final report, the plight of persons displaced from their homes by adverse environmental conditions.[77] Moreover, she used the term *environmental refugee* by rather apodictically stating that *de lege lata*

> [s]uch persons have a wide range of rights – whether or not they qualify as "refugees" under the restrictive definition provided in the 1951 Convention relating to the Status of Refugees or its 1967 Protocol; or under the more expansive definitions in the 1969 Organization of African Unity Convention governing the Specific Aspects of Refugee Problems in Africa or the 1984 Cartagena Declaration on Refugees.[78]

70 *Ibid.*, para. 137(h).
71 Renamed by ECOSOC in 1999 as Sub-Commission on the Promotion and Protection of Human Rights, see also www2.ohchr.org/english/bodies/subcom/index.htm.
72 E/CN.4/Sub.2/1994/9, para. 6.
73 *Ibid.*, para. 7.
74 *Ibid.*, para. 53.
75 *Ibid.*, paras. 38 and 145.
76 *Ibid.*, paras. 82, 89 and 124.
77 *Ibid.*, para. 154; E/CN.4/Sub.2/1993/7, paras. 101–106.
78 *Ibid.*

She further stressed that regardless of this normative-legal discussion, in the end at a minimum such persons have the right to life, the right to health, food and shelter, and the right not to be sent to any location where their lives or security are endangered.[79]

Without deepening a notional-normative discussion, Ksentini defined and clarified at the same time that "For the purposes of this report, the Special Rapporteur uses 'environmental refugee' to refer to anyone forced to leave his or her normal habitat because of serious environmental disruption. This includes those who flee their homes temporarily and those forced to flee permanently internally or across international borders."[80]

It seems, however, that in light of the circumstances, significance and authoritative character of the mandating and approving institutions and organs/bodies as well as the teleology of this study series undertaking, the terminology used and its description appear – rather as an attempt – to offer a legal definition of *environmental refugee* with some legal and temporal validity.

The definition is followed by a typology of environmental flight reasons,[81] before the Special Rapporteur calls on the international community in the instant case to implement the actionable suggestions made by Francis Deng, the then Representative of the Secretary-General on internally displaced persons, who addressed the protection needs and imperatives of environment-induced internally displaced persons.[82]

Finally, Ksentini addressed the UNHCR as the lead agency mandated to "facilitate the delivery of essential services to displaced persons in emergency situations, on humanitarian grounds,"[83] as she put it in her final report, which appears again as a deterrent, since environment-induced displacement as such does not belong to the UNHCR's protection portfolio – as it is still valid. The Special Rapporteur ended on this point by recognizing the UNHCR's acknowledgment to deepen its understanding of the relationship between migration/refugee flows and development and environment issues.[84]

This was particularly important as causes for "massive displacement of peoples," like drought and desertification, were discussed. There was a need to anticipate solutions in view of global protection needs of environmental- and climate-induced displaced persons, as those generally most affected and least protected.[85]

79 *Ibid.*
80 *Ibid.*, para. 155.
81 *Ibid.*, paras. 156–158.
82 E/CN.4/1993/35, annex and E/CN.4/1994/44 and Add.1.
83 E/CN.4/Sub.2/1994/9, para. 160.
84 *Ibid.*, para. 160; see also S. Aga Khan, Legal Problems Relating to Refugees and Displaced Persons, 1976; A/AC.96/199, UNHCR, Note on International Protection submitted to the Executive Committee of the High Commissioner's Programme, 1992; see also UNHCR, statement at UNCED, Rio de Janeiro, June 10, 1992, *cf.* D. Miserez, Prince Sadruddin Aga Khan, Humaniste et Visionnaire, 2017, p. 183 *et seq.*
85 E/CN.4/Sub.2/1994/9, paras. 178 and 249.

Similarly, whereas the social and human dimensions of climate change as one of the most important drivers of ecosystem changes have gained broad recognition over the past few years, the nexus between climate change and the effective enjoyment of human rights has, however, only recently been acknowledged.[86]

In summary, over the years, the Special Rapporteur of the Sub-Commission on Human Rights and the Environment produced a series of reports, starting with the preliminary report in 1991 and two progress reports in 1992 and 1993,[87] before submitting her final report. These concatenated studies need to be read in conjunction with each other if one is to observe that the issue of climate displacement had no primacy before the second progress report, which then resulted in the submission of the Rapporteur's final report, along with a number of recommendations and guidelines, in conformity with the Sub-Commission's request, as approved by the Commission on Human Rights on July 6, 1994, in Geneva.

Ksentini's final report was the culmination of a study process initiated in 1989, when several NGOs submitted information and evidence to the Sub-Commission on the need for all UN human rights bodies to study further the interdependences and mutual interactions between human rights and the environment. It is also to a large extent the merit of NGOs to have advanced the concerns over *environmental refugees*.

## 2 *The Right to Be Cold*[88] *and Not Displaced: The Inuit Petition*

In December 2005, Sheila Watt-Cloutier, a representative of the Canadian Inuit People who at the time chaired the Inuit Circumpolar Conference (ICC),[89] filed what would become known as the Inuit Petition with the Inter-American Commission on Human Rights.[90] The petition claimed the responsibility of the United States for the effects of global warming, through its acts and omissions, which it alleged constitute a violation of the rights of the Inuit. Although rejected, the petition is an important acknowledgment of the human rights climate change nexus, notably because the Inter-American Commission later invited the Inuit Alliance, in February 2007, to testify on the linkages between global warming and the rights of the Inuit.[91]

86 See J. Knox, Climate Change and Human Rights Law, 2009, p. 190 *et subseq.*

87 E/CN.4/Sub.2/1991/8; E/CN.4/Sub.2/1992/7; E/CN.4/Sub.2/1993/7.

88 S. Watt-Cloutier, The Right to Be Cold: One Woman's Story of Protecting Her Culture, the Arctic and the Whole Planet, 2016.

89 The Inuit are a linguistic and cultural group descended from the Thule people whose traditional range spans four countries (i.e. Chukotka in the Russian Federation, northern and western Alaska in the United States, northern Canada and Greenland). See Petition to the Inter-American Commission on Human Rights seeking relief from violations resulting from global warming caused by acts and omissions of the United States, December 7, 2015.

90 ICC petition seeking relief from violations resulting from global warming caused by acts and omissions of the United States, December 7, 2005.

91 Letter from the Assistant Executive Secretary of the Inter-American Commission on Human Rights (IACHR) of the Organization of American States (OAS) to the Inuit Alliance, Earthjustice and the Centre for International Environmental Law of February 1, 2007.

The initial impulse for the petition came from a cognition that the Artic and the information it contains are central to society and the international community. A study commissioned by the Artic Council in 2004 on the impacts of a warming Artic starts with the recognition that "[t]he Artic is of special importance to the world and it is changing rapidly. It is thus essential that decision makers have the latest and best information available regarding ongoing changes in the Artic."[92] This study was essential as it made scientific findings accessible to policy-makers and beyond to a larger audience. Among its key findings was the conclusion that Arctic vegetation zones were very likely to shift, causing wide-ranging impacts sometimes resulting in the loss of habitat, displacement of species and in delayed migration of new species from the south, which would in turn lead to a loss of present biodiversity.[93]

Against the background of these threats for the livelihood of the Inuit people, who were at the risk of losing the basis of their existence, they decided to act and directly filed the so-called Inuit Petition (i.e. without exhaustion of domestic remedies) with the support of the ICC, which represents sixty-two Inuit peoples in Canada and the United States of America.[94] The petitioners requested that the Inter-American Commission engage in an investigation, inquiry and public hearing to ascertain the liability and responsibility.[95] Furthermore, the petition also requested the relief of actions taken by the United States to facilitate the Inuit peoples' adaptation to climate change.

The rights infringed according to the petition included the right to health and life, and the right to property and land, as well as the right to freedom of movement and shelter. The line of argument set out in the petition was to claim an infringement of the right to life under Article 1 of the American Declaration of the Rights and Duties of Man owing to the melting of permafrost as a result of global warming, which would generate the loss of the Inuits' livelihoods. The decrease of species would inevitably generate food scarcity leading thus to food insecurity. Also emphasized was the negative impact on the Inuit people's traditional and cultural lifestyle. Furthermore, the melting of permafrost also entails an infringement of property

92 S. J. Hassol, Arctic Climate Impact Assessment (ACIA), 2004. NB: The Arctic Council commissioning this assessment then charged two of its working groups, the Arctic Monitoring and Assessment Programme (AMAP) and the Conservation of Arctic Flora and Fauna (CAFF), along with the International Arctic Science Committee (IASC), with its implementation.

93 S. J. Hassol, Arctic Climate Impact Assessment (ACIA), 2004, p. 52.

94 The ICC was established in 1973 and construed as a non-profit transnational organization for Inuit people. The goal of the organization is to promote Inuit people's welfare and sustainability. See www.inuitcircumpolar.com (accessed March 5, 2019).

95 The Inuit people directly filed the petition with the Inter-American Commission on Human Rights without exhausting local (legal) remedies because the petitioner claimed the exception of such procedure under Art. 31.2(a) of the Rules of Procedure of the Inter-American Commission on Human Rights. The petition claimed that the exhaustion requirement "[s]hall not apply when [. . .] the domestic legislation of the State concerned does not afford due process of law for protection of the rights or rights that have allegedly been violated." See ICC petition seeking relief from violations resulting from global warming caused by acts and omissions of the United States, December 7, 2005, p. 111.

rights under Article 23 of the American Declaration of the Rights and Duties of Man. Housing, land and property rights protection also cover customary land rights and rights in rem of the Inuit people, as they were losing their means of subsistence that were adapted for this particular environment such as their fishing rights. The melting of permafrost and the changes to the natural climatic environment thus violated their rights to life, health, culture and free movement.

The Inter-American Commission rejected the petition in November 2006, and the given reasoning was that the information the petition provided was insufficient to ascertain a responsibility and liability. Yet, while the petition itself failed, the attempt to link climate change and human rights nevertheless succeeded, since the Commission agreed to hold an information hearing in March 2007 with a view to investigating the link between global warming and human rights and invited the petitioners to testify on this relationship.[96]

It is also important to put the petition, and its rejection, in the larger context, in order to understand its significance and what achievement this constituted. At the time when the petition was filed, there was no scientific affirmation of the link between global warming and pollution. It was only in February 2007 that the IPCC concluded that human pollution was very likely the cause of global warming, which allowed the petitioners to request this hearing of the Inter-American Commission dedicated to this matter, which it was hoped would help the Commission to progressively develop its jurisprudence in future similar cases.

### 3 *Landmark Recognition: The Malé Declaration*

The Malé Declaration,[97] adopted at the Small Island Developing States (SIDS) Conference convened by the Maldives on November 13–14, 2007, constitutes the first explicit recognition of the impact of climate change on human rights at the international level. The Declaration stresses that the environment provides the infrastructure for human civilization and that climate change poses the most immediate, fundamental and far-reaching global threat to the environment, and thus to individuals and to communities. Most importantly, it asserts and articulates key human rights protection apprehensions as States expressed their concerns "that climate change has clear and immediate implications for the full enjoyment of human rights including inter alia the right to life, the right to take part in cultural life, the right to use and enjoy property, the right to an adequate standard of living,

96 See Earthjustice, Global Warming and Human Rights gets Hearing on the World Stage, Earthjustice Managing Attorney Martin Wagner gives Testimony Before Inter-American Commission on Human Rights. See also Inuit leader and activist Sheila Watt-Cloutier's testimony before the Inter-American Commission on Human Rights, March 1, 2007.

97 AG/RES.2429 (XXXVIII-O/08), Human Rights and Climate Change in the Americas; Malé Declaration on the Human Dimension of Global Climate Change, November 14, 2007.

the right to food, and the right to the highest attainable standard of physical and mental health."[98]

The Malé Declaration is of particular significance as it instigated a global policy process by virtue of bringing the foundational principles of human rights into the various climate change discussions as a reference value. It also urged the cooperation of the OHCHR and the UN Human Rights Council in assessing the human rights implications of climate change.[99] More specifically, in its operative paragraph 4, the Declaration requests "[t]he Office of the United Nations High Commissioner for Human Rights to conduct a detailed study into the effects of climate change on the full enjoyment of human rights, which includes relevant conclusions and recommendations thereon, to be submitted prior to the tenth session of the Human Rights Council."[100] Additionally, in the ensuing operative paragraph 5, it requests the UN Human Rights Council to convene, in March 2009, a debate on human rights and climate change. Concluding, the Malé Declaration advocates for a changed approach to climate change stressing the commitment of SIDS to a dialogue process that is not merely people-centered and rights-based, but also inclusive. Attempting to ensure a human rights primed and collaborative climate change debate, the Declaration states that it is: "Committed to an inclusive process that puts people, their prosperity, homes, survival and rights at the centre of the climate change debate."[101]

**Malé Declaration on the Human Dimension of Global Climate Change**
We the representatives of the Small Island Developing States ... solemnly request:

1. The international community to commit in Bali to a formal process that will ensure a post-2012 consensus to protect people, planet and prosperity by taking urgent action to stabilize the global climate and ensure that temperature rises fall well below 2°C above pre-industrial averages, and that greenhouse gas concentrations are less than 450ppm, consistent with the principles of common but differentiated responsibilities.
2. The members of AOSIS [Alliance of Small Island States] in New York to consider including the human dimension of global climate change as one of the agenda items for the meeting of AOSIS Ministers in Bali, and to

98 Malé Declaration on the Human Dimension of Global Climate Change, November 14, 2007, preambular para. 13.

99 AG/RES.2429 (XXXVIII-O/08), Human Rights and Climate Change in the Americas; Malé Declaration on the Human Dimension of Global Climate Change, November 14, 2007.

100 Malé Declaration on the Human Dimension of Global Climate Change, November 14, 2007, preambular para. 13.

101 *Ibid.*

**(*cont.*)**

explore possible alternatives for advancing this initiative in Bali in order to stress the moral and ethical imperatives for action.

3. The Conference of the Parties of the United Nations Framework Convention on Climate Change, with the help of the Secretariat, under article 7.2(l), to seek the cooperation of the Office of the United Nations High Commissioner for Human Rights and the United Nations Human Rights Council in assessing the human rights implications of climate change.
4. The Office of the United Nations High Commissioner for Human Rights to conduct a detailed study into the effects of climate change on the full enjoyment of human rights, which includes relevant conclusions and recommendations thereon, to be submitted prior to the tenth session of the Human Rights Council.
5. The United Nations Human Rights Council to convene, in March 2009, a debate on human rights and climate change.

Committed to an inclusive process that puts people, their prosperity, homes, survival and rights at the centre of the climate change debate, other AOSIS members not present in Malé are invited to endorse this Declaration.

*Malé, Republic of Maldives, November 14, 2007*

It is important to emphasize that the SIDS Conference does not belong to the UN system and that it therefore does not have the competence to mandate any UN body or department, such as OHCHR or the Human Rights Council. As will be demonstrated later, these requests were important signals however to the UN human rights machinery and through their active engagement in the Human Rights Council, SIDS, notably the Maldives, garnered support for a resolution in the Council formalizing the requests initially formulated in the Declaration. However, in spite of the progress made, the human rights proviso is not systematically included in the climate change debates and processes. The recent statement by Mary Robinson, former President of Ireland and UN High Commissioner for Human Rights, who currently leads her own foundation dedicated to climate justice, at a UN Expert Meeting on Climate Change and Human Rights about the implementation of the Paris Agreement attests to this effect, as she emphasized the importance to ensuring a human rights-based approach at the operational level, stressing that bringing "human rights sensitive approaches in 'after the fact' " would not be possible. She urged that State delegates and other parties needed to see the linkages now, as the window of opportunity was about to close.[102]

102 Mary Robinson, Opening of the OHCHR Expert Meeting on Climate Change and Human Rights, October 6, 2016.

In addition to the longer-term impact of the Malé Declaration in the rapprochement of the human rights and climate change tracks, it also had an immediate effect, since only a month later, the then Deputy High Commissioner for Human Rights, Kyung-wha Kang, addressed the Bali Climate Change Conference (COP 13). In her statement, she reminded States of their human rights obligations with regard to climate change and stressed that "[t]he human rights approach compels us to look at the people whose lives are most adversely affected and to urge governments to integrate their human rights obligations into policies and programs to deal with the climate change as well as to the international community to assist in this process."[103] Kang further outlined the obligations international human rights law imposes on States that are relevant to addressing human vulnerabilities to climate change, which "include the obligation to provide better housing, located away from hazardous zones; improved access to sanitation, safe-drinking water and healthcare; access to adequate food; effective participation in planning and decision-making; accountability; as well as access to information and justice."[104]

## III MOBILIZING THE UNITED NATIONS HUMAN RIGHTS MACHINERY

### 1 *Sparking the Discussions: 2009 High Commissioner's Report*

The Malé Declaration set into motion a process that culminated in the adoption by the Human Rights Council of resolution 7/23 on human rights and climate change in March 2008. With this resolution, the implications of climate change for the enjoyment of human rights were ascertained for the first time at the level of the Human Rights Council. The resolution also commissioned the High Commissioner for Human Rights (i.e. the UN official with principal responsibility for UN human rights activities as per General Assembly resolution 48/141)[105] to conduct a study on States' obligations under international human rights law to protect those rights from the effects of climate change.[106]

The report of the UN High Commissioner, which was presented a year later at the 10th session of the Human Rights Council in March 2009, concluded that climate change had a range of implications for the effective enjoyment of human rights, be it of a direct nature or indirectly, as the impact of climate change on human rights may often translate through discriminatory policies and treatment. Most importantly, the report also outlined the different legal guarantees for those displaced as a result of climate change, whereby it differentiated between different forms of displacement:

103 Kyung-wha Kang, Deputy High Commissioner for Human Rights, address on Climate Change and Human Rights, Conference of the Parties to the UNFCCC and its Kyoto Protocol, December 3–14, 2007, Bali, Indonesia.

104 *Ibid.*

105 A/RES/48/141 of December 20, 1993; see also B. G. Ramcharan, The United Nations High Commissioner for Human Rights, 2002, p. 73 *et seq.*

106 A/HRC/RES/7/23 of March 28, 2008.

Persons affected by displacement within national borders are entitled to the full range of human rights guarantees by a given State, including protection against arbitrary or forced displacement and rights related to housing and property restitution for displaced persons. To the extent that movement has been forced, persons would also qualify for increased assistance and protection as a vulnerable group in accordance with the Guiding Principles on Internal Displacement.[107]

While the negative effects of climate change on the realization of human rights are not contested, the key question remains as to whether this impact would qualify as a human rights violation itself.[108] Hence, whereas the report concludes that climate change poses a threat to the enjoyment of human rights, it does not affirm that climate change itself is a violation of human rights.

**OHCHR Key Messages on Human Rights and Climate Change**

In order to foster policy coherence and help ensure that climate change mitigation and adaptation efforts are adequate, sufficiently ambitious, non-discriminatory and otherwise compliant with human rights obligations, the following considerations should be reflected in all climate action:

1. To mitigate climate change and to prevent its negative human rights impacts
2. To ensure that all persons have the necessary capacity to adapt to climate change
3. To ensure accountability and effective remedy for human rights harms caused by climate change
4. To mobilize maximum available resources for sustainable, human rights-based development
5. International cooperation
6. To ensure equity in climate action
7. To guarantee that everyone enjoys the benefits of science and its applications
8. To protect human rights from business harms
9. To guarantee equality and non-discrimination
10. To ensure meaningful and informed participation

*www.ohchr.org/Documents/Issues/ClimateChange/KeyMessages_on_HR_CC.pdf*

The report also identifies one of the key challenges in relation to litigation (i.e. the attribution of climate change-related harm to acts or omissions of States), which is

107 A/HRC/10/61, para. 57.

108 *Ibid.*, para. 69; see also E. Cameron, Human Rights and Climate Change: Moving from an Intrinsic to an Instrumental Approach, 2010, p. 701 *et subseq.*

often cited as the reason for which the human rights framework would only bring limited protection to climate displaced persons. The report recognized the need for redress and adequate legal protection against climate change-related or -induced violations, yet affirms the need to further study protection mechanisms for climate change displaced persons.[109] In terms of protection, notwithstanding these questions, the human rights obligations of States provide the affected individuals with important guarantees.[110]

With this report, the OHCHR was the first UN entity to examine and recognize the nexus between climate change and the effective enjoyment of human rights.

### 2 *Climate Change on the Human Rights Council's Agenda*

The Human Rights Council is the principal intergovernmental body dealing with human rights in the United Nations.[111] Its resolutions and decisions are not legally binding but have normative value and contain strong political commitments, similar to those of the General Assembly. The Council is responsible for promoting universal respect for the protection of all human rights and fundamental freedoms for all, without distinction of any kind and in a fair and equal manner.[112]

An important and the most mediatized aspect of the Council's mandate is that it addresses human rights situations in specific countries, for instance by establishing country-specific special rapporteurs, international commissions of inquiry or fact-finding missions, or through the holding of special sessions. The Council also analyzes specific thematic human rights issues, covering a plethora of topics at each session relating to the full set of human rights (i.e. civil and political as well as economic, social and cultural rights).[113]

Another important element of the Council's work is setting international human rights standards. This had been the initial focus of the work of its predecessor, the Commission on Human Rights, and has successfully been continued by the Council. The Council for instance elaborated new international conventions relating to disappearances and to disability, optional protocols to the International Covenant on Economic, Social and Cultural Rights and to the Convention on the Rights of the Child, and international standards on a variety of themes, such as indigenous rights

109 A/HRC/10/61; see also M. S. Chapman, Climate Change and the Regional Human Rights Systems, 2009–2010, p. 37 *et seq.*; A. Boyle, Human Rights and the Environment: Where Next?, 2012, p. 613 *et subseq.*; Council of Europe, Manual on Human Rights and the Environment, 2012; P. Stephens, Applying Human Rights Norms to Climate Change: The Elusive Remedy, 2010, p. 49 *et seq.*

110 A/HRC/10/61, para. 71.

111 A/RES/60/251 of March 15, 2006; W. Kälin, Turning the Vision of an Effective Human Rights Council into Reality, 2006, p. 225 *et seq.*; A/59/565, para. 283; A/RES/60/1 of October 24, 2005, para. 157.

112 A/RES/60/251 of March 15, 2006, operative para. 2.

113 R. Freedman, The United Nations Human Rights Council: A Critique and Early Assessment, 2013, p. 71 *et seq.*; see also B. G. Ramcharan, The Protection Role of UN Human Rights Special Procedures, 2009.

(by virtue of elaborating a declaration) among others. Furthermore, the Council promotes human rights through dialogue, including by monitoring capacity-building and by providing as well as facilitating technical assistance.[114]

Over the few years of its existence, the Council has proved to be a dynamic body venturing into many new areas, regularly addressing emerging human rights issues as well as examining the human rights aspects of matters dealt with in other dedicated forums, such as the impact of arms transfers on human rights in armed conflicts.

In the same vein, the Council seized itself of the human rights and climate change agenda further to the adoption of the Malé Declaration. It acknowledged the interrelationship between the enjoyment of human rights and climate change by resolution 7/23, which had mandated the High Commissioner's report on the issue, and again by resolution 10/4 adopted on March 25, 2009, further to the consideration of the said report.[115] Resolution 10/4 noted that "climate change-related impacts have a range of implications, both direct and indirect, for the effective enjoyment of human rights,"[116] and that such effects "will be felt most acutely by those segments of the population who are already in a vulnerable situation."[117] Further to this resolution, a panel discussion on the relationship between climate change and human rights was held at the 11th session of the Council to improve the understanding of the implications of climate change on the full enjoyment of human rights, and to discuss the implications on policy-making and further development of international law.

In September 2011, the Council adopted a further resolution. By resolution 18/22, it affirmed that "human rights obligations, standards, and principles have the potential to inform and strengthen international and national policy-making in the area of climate change, promoting policy coherence, legitimacy, and sustainable outcomes,"[118] and the Council reiterated "its concern that climate change poses an immediate and far-reaching threat to people and communities around the world and has adverse implications for the full enjoyment of human rights."[119] The Council therein also requested OHCHR to convene a seminar to address the adverse impacts of climate change on the full enjoyment of human rights and to prepare a summary report of the seminar.[120]

114 R. Freedman, The United Nations Human Rights Council: A Critique and Early Assessment, 2013, p. 77; B. G. Ramcharan, The Protection Role of UN Human Rights Special Procedures, 2009; A/RES/66/138 of December 19, 2011; A/RES/63/117 of December 10, 2008; A/RES/61/295 of September 13, 2007.

115 A/HRC/RES/10/4 of March 25, 2009.

116 *Ibid.*, preambular para. 7.

117 *Ibid.*, preambular para. 8.

118 A/HRC/RES/18/22 of September 30, 2011, preambular para. 17.

119 *Ibid.*, operative para. 1.

120 *Ibid.*, operative para. 2.

The seminar, in implementation of the resolution, was held in February 2012 with a view to following up on the call for respecting human rights in all climate change-related actions and policies and forging stronger cooperation between the human rights and climate change communities. The then UN High Commissioner for Human Rights, Navi Pillay, emphasized in a statement that addressing the issue "requires considering how to internalize the human consequences of climate change from a rights-based perspective, rather than limiting ourselves to quantitative dimensions alone."[121]

It took a further three years for the Council to come to an agreement on another resolution on the issue of climate change. A resolution had been negotiated in 2013, but since it was not possible to reach a consensus on a number of fronts, even though several rounds of informal consultations were held with all States and various stakeholders, the main sponsors of the draft resolution ultimately decided not to table their text for adoption by the Council. In June 2014, the sponsors submitted a new draft proposal, which was eventually adopted to become Council resolution 26/27.[122]

The Council with this resolution marked the beginning of the consideration of climate change annually in the context of a dedicated full-day discussion. The Council decided that the thematic focus of the first full-day discussion would be (i) on identifying challenges and ways forward toward the realization of human rights for all, as well as the measures and best practices to promote and protect human rights that can be adopted by States in addressing the adverse effects of climate change on the full and effective enjoyment of human rights, and (ii) on how climate change has had an adverse impact on States' efforts to progressively realize the right to food, and policies, lessons learned and good practices.[123] Moreover, the resolution made a strong appeal for cooperation, calling

> upon all States to continue to enhance international dialogue and cooperation in relation to the adverse impact of climate change on the enjoyment of human rights, including the right to development, particularly in developing countries, especially least developed countries, small island developing States and African countries, including through dialogue and measures, such as the implementation of practical steps to promote and facilitate capacity-building, financial resources and technology transfer.[124]

The following year, in July 2015, the Council adopted resolution 29/15, which this time focused on the adverse impact of climate change on States' efforts to progressively realize the right of everyone to the enjoyment of the highest attainable

121 A/HRC/20/7, para. 64; see also DPI/2483, OHCHR, United Nations Joint Press Kit for the Bali Climate Change Conference, December 3–14, 2007: The Human Rights Impact of Climate Change, November 2007.

122 A/HRC/RES/26/27 of June 27, 2014, operative para. 6.

123 *Ibid.*

124 *Ibid.*, operative para. 5.

standard of physical and mental health. It mandated a panel discussion on this topic with a view to identifying lessons learned and good practices, as well as a detailed analytical study on the relationship between climate change and the human right of everyone to the enjoyment of the highest attainable standard of physical and mental health.[125]

The subsequent July 2016 resolution 32/33 of the Council clearly reflects the Paris *acquis*. The resolution obviously welcomed the Paris Agreement as it acknowledged that "parties should, when taking action to address climate change, respect, promote and consider their respective obligations on human rights, the right to health, the rights of indigenous peoples, local communities, migrants, children, persons with disabilities and people in vulnerable situations and the right to development, as well as gender equality, empowerment of women and intergenerational equity,"[126] and therefore established beyond any doubt that climate change was indeed a human rights concern. This new initiative contained some more, largely positive elements, such as a reference to the Geneva Pledge on Human Rights in Climate Action, launched in February 2015 by Costa Rica, a voluntary pledge affirming the importance of addressing the human rights implications of climate change and the need for human rights to inform responses to the climate change challenges,[127] and to the Sendai Framework for Disaster Risk Reduction 2015–2030, as well as to climate justice and improved cooperation. The thematic focus of this annual resolution was on the adverse impact of climate change on the rights of the child. It mandated a panel discussion to be held on this topic as well as an analytical study.[128] Of note is furthermore that in the resolution the Council recognized that "both sudden-onset natural disasters and slow-onset events [. . .] have adverse effects on the full enjoyment of all human rights."[129]

Other resolutions within the Human Rights Council that dealt with the broader issue of human rights and the environment and that also touched upon the impact of climate change on human rights are resolutions 16/11,[130] 19/10,[131] 25/21[132] and 28/11.[133]

At face value, this link between the enjoyment of human rights and climate change would appear evident, given the impact that environmental degradation is projected to have on the lives of people and therefore on their enjoyment of their rights. The negotiations of the various resolutions proved far from being consensual however and continues to remain a sensitive matter, particularly when it comes to

125 A/HRC/RES/29/15 of July 2, 2015, operative paras. 3 and 4.
126 A/HRC/RES/32/33 of July 1, 2016, preambular para. 17.
127 For further information on the Geneva Pledge, see http://climaterights.org/our-work/unfccc/geneva-pledge/ (accessed May 10, 2019).
128 A/HRC/RES/32/33 of July 1, 2016, operative paras. 4 and 5.
129 *Ibid.*, operative para. 1.
130 A/HRC/RES/16/11 of March 24, 2011.
131 A/HRC/RES/19/10 of March 22, 2012.
132 A/HRC/RES/25/21 of March 28, 2014.
133 A/HRC/RES/28/11 of March 26, 2015.

references to the climate change framework and the climate negotiation process, which some considered a separate track in which human rights forums should not interfere. There were in fact a number of voices throughout the various processes arguing that the Human Rights Council should not seize itself of issues that are not human rights issues *stricto sensu*, but rather extraneous to its agenda and that climate change issues should be discussed within the appropriate framework (i.e. the climate change framework).[134]

Particular concerns are therefore typically expressed by the countries of the Western European and Others Group (WEOG) on the references to the climate change negotiations and the UNFCCC. While others, notably those most affected by climate change, as well as NGOs, continue to push for closer linkages between the climate change and the human rights frameworks, advocating for an active role of the Council in this regard.

The divide became particularly evident during the unsuccessful 2013 negotiations of a draft text. The draft resolution that had been prepared by Bangladesh and the Philippines built on previous Council resolutions, in particular resolution 18/22, but in addition incorporated many elements from climate change frameworks, especially the UNFCCC. This gave rise to a number of concerns that the United States and the European Union raised, who felt that the human rights focus of the resolution was being diluted by the climate change language. They therefore requested the deletion of many of the preambular paragraphs taken from the UNFCCC or climate change frameworks. The United States also did not support the inclusion of a list of concrete human rights affected by climate change, since it felt this was selective. There were also concerns that the resolution would prejudge the ongoing climate change negotiations in other fora.[135]

The European Union again reiterated this position at the time of the adoption of resolution 26/27, stating: "We believe it is important to encourage the consideration of human rights principles in the work of the UNFCCC, but the Council must not provide an alternative forum for work being undertaken on addressing climate change in the UNFCCC framework."[136] When discussing Council resolution

[134] See, e.g., Statement of Italy on behalf of the European Union of June 27, 2014, Human Rights and Climate Change, EU Explanation of Position, UN Human Rights Council, 26th session, item 3, resolution L.33/Rev.1; on the question of how the interface between the climate issue and human rights should be framed, see M. Khor, South Centre, A Human Rights Approach to Climate Change, Statement at the annual full-day discussion on specific themes related to human rights and climate change at the 28th session of the Human Rights Council, March 6, 2015; and M. Robinson, Statement at the annual full-day discussion on specific themes related to human rights and climate change at the 28th session of the Human Rights Council, March 6, 2015.

[135] J. McAdam, M. Limon, Human Rights, Climate Change and Cross-Border Displacement: The Role of the International Human Rights Community in Contributing to Effective and Just Solutions, 2015, p. 11; Notes of the 2013 informal consultations on the draft resolution, on file with the author.

[136] Statement of Italy on behalf of the European Union of June 27, 2014, Human Rights and Climate Change, EU Explanation of Position, UN Human Rights Council, 26th session, item 3, resolution L.33/Rev.1.

29/15, India by contrast stated that the UNFCCC was the appropriate venue for the topic and any human rights discourse on climate change should complement the convention and underline people's fundamental right to development. This was somehow echoed by the European Union, the United States, Japan and Australia, even though in different terms, at various points during the consultations on resolution 32/33, expressing concerns about the Council overstepping its mandate with this resolution, and indicating in some places that they thought the UNFCCC was a more appropriate forum for some of the language in the draft resolution. This clearly demonstrates the intricacies surrounding the debate on the nexus between climate change and human rights which, in spite of a growing recognition of the interrelations, especially when it comes to concrete obligations arising from human rights.

## 3 *The Human Rights Machinery in Action: The Social Forum*

Another body within the human rights machinery, which fulfills an essential forum function, the Social Forum, also addressed the issue of human rights and climate change at one of its sessions.

The Social Forum is a subsidiary organ of the Human Rights Council and was preserved in the context of the transition from the Commission on Human Rights to the Human Rights Council by resolution 6/13, which was adopted on September 27, 2008, without a vote.[137] The rationale for its initial establishment was growing concern at the effects of globalization on the enjoyment of human rights, and it was to specifically focus on economic, social and cultural rights. The Forum was first created by the former Sub-Commission on the Promotion and Protection of Human Rights by resolution 2001/24 of August 17, 2001. The Council, when it replaced the Commission for Human Rights, reaffirmed that the Social Forum was "a unique space for interactive dialogue between the United Nations human rights machinery and various stakeholders, including grass-roots organizations, and underlines the importance of coordinated efforts at national, regional and international levels for the promotion of social cohesion based on the principles of social justice, equity and solidarity as well as to address the social dimension and challenges of the ongoing globalization process."[138]

The Forum owes its uniqueness to the fact that it is open to a wide array of different stakeholders, many of which would not be allowed to participate in the meetings of the UN human rights machinery, which is largely – by definition – of an intergovernmental nature and where in fact the most salient human rights issues are debated and decisions taken on them. The Council resolution explicitly determines the wide-ranging participation as it indicates that the Forum would, besides State

137 A/HRC/RES/6/13 of September 27, 2008.
138 *Ibid.*, operative para. 3.

representatives, be open for participation for "all other interested stakeholders," listing for instance "newly emerging actors such as small groups and rural and urban associations from the North and the South, anti-poverty groups, peasants' and farmers' organizations and their national and international associations, voluntary organizations, youth associations, community organizations, trade unions and associations of workers, as well as representatives of the private sector."[139] The debates in the Forum are therefore an important complement to the predominantly State-driven agenda of the Human Rights Council to exchanging views across a different and more diverse spectrum of stakeholders in order to formulate concrete recommendations to address today's human rights challenges and getting beyond some of the more political (including domestic) considerations.

In 2010, the Social Forum was dedicated to discussing the measures and actions to address the impact of climate change on the full enjoyment of human rights at the local, national, regional and international levels as mandated by Human Rights Council resolution 13/17.[140] The Forum concluded, in line with the Cancún Adaptation Framework (CAF), that the movement of people should be seen as a possible and legitimate adaptation strategy, among other options, if not the only one at a certain point. It recommended in this regard that

> the situation of persons displaced as a result of climate change be addressed, and that national Governments in disaster-prone countries invest in disaster risk reduction planning, mechanisms and procedures, as well as in other adaptation measures in the face of the already present negative impacts of climate change, and that the members of the international community increase their efforts in responding to international disasters and in investing in preparedness, adaptation and mitigation through sustainable development options, which may imply the transfer of best available technologies.[141]

Similar to the Sub-Commission in 1994,[142] it also recommended the creation of a new special procedures mandate.[143]

## 4 *Instilling Independent Expertise into the Debate*

Another track followed in the Human Rights Council focused on the issue of human rights and the environment. This process eventually culminated in the creation of the mandate of the independent expert on human rights and the environment in March 2012.[144] While Ksentini's report bore a similar title, the mandate conferred to her originated in the Sub-Commission, an expert body, and was limited to the

139 *Ibid.*, operative para. 10.
140 A/HRC/RES/13/17 of March 25, 2010, operative para. 5.
141 A/HRC/16/62, para. 60(f).
142 E/CN.4/Sub.2/1994/9.
143 A/HRC/16/62.
144 A/HRC/RES/19/10 of March 22, 2012.

preparation of a research study. This new special procedures mandate, as established by the Human Rights Council, is of very different nature, and its creation has been a milestone in the Council's engagement on environmental issues, including climate change. Independent experts or special rapporteurs, while being mechanisms of the Human Rights Council, enjoy independence, and have the crucial role to examine a particular thematic issue contributing to the progressive development of international law, or to investigate human rights violations in a particular country and publicly report on it.[145] They are normally appointed for a period of three years, with the possibility of one extension, and they are in fact the main source of independent – nonpartisan – expertise provided to the Council (aside from NGOs) and serve as the basis for the Council's decisions. As a matter of fact, Human Rights Council resolutions regularly draw on the reports of special rapporteurs or independent experts, and include specific language they have used, particularly in those instances where States seek to introduce new language that has not yet been negotiated and has thus not yet been agreed upon, and which they would like to be corroborated by an independent source outside their immediate sphere of influence. One of the specificities of special rapporteurs is that they have the competence to carry out fact-finding missions to countries, participate in academic consultations or other intergovernmental processes and engage with a broad range of actors worldwide.

The process leading up to the mandate creation proved to be quite divisive. Diverging points of view within the original core group of sponsors of the climate change agenda in the Human Rights Council (i.e. Bangladesh, Costa Rica, the Maldives, Morocco, the Philippines, Slovenia and Switzerland) on whether the new special procedure mandate should focus more narrowly on climate change or on the environment in general, proved to be irreconcilable and as a result, Bangladesh and the Philippines left the core group and continued to push for their own resolution on climate change and human rights. This also explains why in 2013 the focus of the climate change resolutions shifted to include more UNFCCC language and why it was getting increasingly difficult to rally States behind the proposals, resulting in a stalemate leaving Bangladesh and the Philippines to renounce to the tabling of their draft resolution in 2013.

The legislative history of the mandate started with the former Sub-Commission on Prevention of Discrimination and Protection of Minorities recommending the creation of a new mandate of a special rapporteur on the issue of climate change and human rights. In 2010, the Social Forum reiterated this recommendation with the support of participating stakeholders, notably NGOs, and it again emerged as one of

[145] O. de Schutter, International Human Rights Law: Cases, Materials, Commentary, 2014, p. 962 *et subseq.*; H. C. Rivera, The United Nations Human Rights Council: Remarks on Its History, Procedures, Challenges and Perspectives, 2015, p. 7 *et subseq.* For the full list of special procedures, see A/HRC/31/39, Annex XI.

the core action points during the seminar organized in February 2012 in the implementation of Council resolution 18/22 on the issue of addressing the adverse impacts of climate change on the full enjoyment of human rights,[146] just the week before the Council's main session in March 2012. At the beginning of the Council's session, during which the mandate was eventually created, in the context of the annual high-level segment at which heads of states and other dignitaries address this principal human rights forum, a number of ministers from concerned countries referred to the creation of such a new mandate on human rights and climate change.[147]

Yet, as mentioned above, the Council followed a different approach with resolution 19/10 on March 22, 2012, when it decided instead to establish the broader mandate on human rights and the environment, to study the human rights obligations relating to the enjoyment of a safe, clean, healthy and sustainable environment, and to promote best practices relating to the use of human rights in environmental policy-making.

Resolution 19/10 sets out the terms of the mandate of the new Independent Expert by requesting the mandate holder inter alia:

> (a) To study, in consultation with Governments, relevant international organizations and intergovernmental bodies, including the United Nations Environment Programme and relevant multilateral environment agreements, human rights mechanisms, local authorities, national human rights institutions, civil society organizations, including those representing indigenous peoples and other persons in vulnerable situations, the private sector and academic institutions, the human rights obligations, including non-discrimination obligations, relating to the enjoyment of a safe, clean, healthy and sustainable environment;
>
> (b) To identify, promote and exchange views on best practices relating to the use of human rights obligations and commitments to inform, support and strengthen environmental policymaking, especially in the area of environmental protection, and, in that regard, to prepare a compendium of best practices;
>
> (c) To make recommendations, consistent with her or his mandate, that could help the realization of the Millennium Development Goals, in particular Goal 7;
>
> (d) To take into account the results of the United Nations Conference on Sustainable Development to be held in June 2012, and to contribute a human rights perspective to follow-up processes.[148]

While not mentioned that explicitly in the mandate's legislative basis, the Independent Expert (now Special Rapporteur) from the inception dealt with the

[146] Such mandate had already been recommended in 1994 by the former Sub-Commission on Prevention of Discrimination and Protection of Minorities; see E/CN.4/Sub.2/1994/9; see also the related discussions at the Social Forum.

[147] See, for instance, the statement of D. Maumoon, State Minister for Foreign Affairs of the Maldives, on March 1, 2012.

[148] A/HRC/RES/19/10 of March 22, 2012, operative para. 2.

question of human rights and climate change, taking into account that some of the key issues in the climate change debate, concerning the human rights nexus, remain the same when it comes to environmental questions at large.[149] In his first report to the Human Rights Council, the new Independent Expert outlined the linkage between substantive and procedural rights and duties in the area of human rights and the environment, such as access to information, public participation and access to legal remedies, and, like the former Representative of the Secretary-General on internally displaced persons Chaloka Beyani,[150] emphasized the key role of compliance with procedural rights in terms of impact on the protection of substantive rights.[151]

The mandate of the Independent Expert was extended in 2015 by Council resolution 28/11.[152] The resolution changed the title from independent expert to special rapporteur and recognized, unlike the previous resolution 19/10, "that climate change, [...] may interfere with the enjoyment of a safe, clean, healthy and sustainable environment."[153] As a result, since the mandate renewal, the Independent Expert focused increasingly on the issue of climate change and dedicated, for instance, one of his annual thematic reports entirely to the human rights obligations relating to climate change.[154] The report outlined the human rights obligations of States relating to climate change, including procedural but also substantive ones.

Other special procedures also significantly contributed to analyzing the linkage between human rights and climate change. In 2011, the Representative of the Secretary-General on internally displaced persons, Walter Kälin, for instance, devoted his report to the General Assembly to climate change and human rights.[155] The Special Rapporteur on adequate housing, Raquel Rolnik, as a component of the right to an adequate standard of living, and on the right to

[149] A/HRC/RES/28/11 of March 26, 2015; Special procedure mandate-holders are either designated as special rapporteurs or independent experts. Special procedures on country situations that are established with the consent of the State concerned are designated as independent experts as opposed to country-specific special rapporteurs, which are normally objected by the country concerned. Since independent experts therefore appear to be less intrusive, States during negotiations of new thematic mandates, sometimes resort to an independent expert rather than a special rapporteur to achieve consensus. Mandates may later be converted during renewal after three years. See M. Limon, T. Piccone, Human Rights Special Procedures: Determinants of Influence, Understanding and Strengthening the Effectiveness of the UN's Independent Human Rights Experts, 2014, p. 15.

[150] A/66/285, para. 81 *et subseq.*

[151] A/HRC/22/43, para. 40 *et seq.* See also Statement of the EU Delegation at the 22nd session of the Human Rights Council delivered during the interactive dialogue with the Independent Expert on the issue of human rights obligations relating to the enjoyment of a safe, clean, healthy and sustainable environment, March 7, 2013; A. Boyle, Human Rights and the Environment: Where Next?, 2012, p. 621 *et subseq.*

[152] A/HRC/RES/28/11 of March 26, 2015.

[153] *Ibid.*, preambular para. 8.

[154] A/HRC/31/52; see also A/HRC/RES/19/10 of March 22, 2012.

[155] A/66/285; A/HRC/13/21.

nondiscrimination in this context not only prepared a report on climate change,[156] but also integrated the climate displacement perspective in a number of other thematic and mission reports.[157] The Special Rapporteur on the human rights of migrants, François Crépeau,[158] and the Special Rapporteur on the right to food, Hilal Elver,[159] also dealt with the issue of climate change.

While the continuous engagement by the Human Rights Council on the issue of climate change reflects the growing political recognition that climate change affects the enjoyment of human rights, there have also been developments in other human rights mechanisms, notably the treaty bodies, which are the guardians of international human rights norms.

In August 2009, the Committee on the Elimination of Discrimination against Women (CEDAW) at its 44th session adopted a statement on gender and climate change, noting that from the examination of State parties' reports, it is apparent that climate change does not affect women and men in the same way and has a gender-differentiated impact. This was an important step forward in advocating for disaster risk reduction responses and the handling of climate displacement to be gender-sensitive.[160]

Acknowledging the different impact of climate change on men and women,[161] CEDAW remained seized of matter and just recently organized a half-day general discussion on gender-related dimensions of disaster risk reduction and climate change on the margins of the 63rd session in February/March 2016. This is an important preparatory step for the elaboration of a General Recommendation on gender-related dimensions of disaster risk reduction and climate change by CEDAW, which is the manner in which the Committee provides guidance to State parties on their human rights obligations in implementing the Convention and particularly on how these relate to the respect, protection and fulfillment of women's human rights in the context of disaster and climate change.[162] The adoption of a General Comment would be a landmark in the recognition of the link between climate change and human rights and the ensuing obligations of States as the Committee would thereby provide its authoritative views on the Convention and contribute to progressively developing international law in the area of climate change.

Concrete guidance by the Committee is needed on how to conceptualize and develop effective measures to address the gender-related dimensions of climate change and disaster risk reduction from a human rights perspective and on how to

156 A/64/255.
157 See A/HRC/10/7/Add.4; A/HRC/13/20/Add.3; see also A/HRC/16/42; A/66/270.
158 A/67/299.
159 A/70/287.
160 CEDAW, Statement of the CEDAW Committee on Gender and Climate Change, 44th session, New York, July 20–August 7, 2009.
161 CEDAW/C/2010/47/GC.1, October 19, 2010.
162 OHCHR, Concept Note for the Half-Day of General Discussion on Gender-Related Dimensions of Disaster Risk Reduction and Climate Change.

implement and monitor them.[163] The adoption of a General Recommendation by CEDAW will also be an important signal for committees of other treaty bodies to follow a similar path.

## IV THE ROAD TO PARIS

### 1 *Creating Momentum*

One of the crucial elements of the mandate of the Special Rapporteur on human rights and the environment is that the resolution requested the mandate holder "contribute to and participate in, where appropriate, intergovernmental conferences and meetings relevant to the mandate."[164] This laid the ground for the active engagement of the independent expert within the climate change framework. One of the specificities of the special rapporteurs, as mentioned earlier, and which distinguishes them fundamentally from most other UN human rights mechanisms, is that their scope of action is not confined to the meeting rooms in Geneva, but that they can actually actively participate in other fora and engage with States and other actors in a variety of ways. Insofar, the creation of this mandate constituted a major step forward in the rapprochement of the climate change and human rights frameworks.[165]

In accordance with this aspect of this mandate, the Special Rapporteur on human rights and the environment participated in the Conference of Parties in Lima (COP 20) as well as in the Climate Change Conference in Bonn, which was held in June 2015. The mandate holder was also joined by many other special rapporteurs on several occasions advocating for the inclusion of human rights in the climate change framework. In an open letter to the State parties to the UNFCCC on the occasion of the meeting of the Ad Hoc Working Group on the Durban Platform for Enhanced Action held on October 20–25, 2014, in Bonn, a large group of special rapporteurs urged "the State Parties to the UNFCCC to recognise the adverse effects of climate change on the enjoyment of human rights, and to adopt urgent and ambitious mitigation and adaptation measures to prevent further harm." The letter continues:

> We call on the State Parties to include language in the 2015 climate agreement that provides that the Parties shall, in all climate change related actions, respect, protect, promote, and fulfil human rights for all. And we urge the State Parties at COP 20 in Lima to launch a work program to ensure that human rights are integrated into all aspects of climate actions.[166]

163 *Ibid.*
164 A/HRC/RES/28/11 of March 26, 2015, operative para. 5(e).
165 B. G. Ramcharan, The United Nations High Commissioner for Human Rights, 2002, p. 57 *et seq.*; B. G. Ramcharan, The Protection Role of UN Human Rights Special Procedures, 2009.
166 Open Letter from Special Procedures mandate-holders of the HRC to the State Parties to the UNFCCC on the occasion of the meeting of the Ad Hoc Working Group on the Durban Platform for Enhanced Action in Bonn, October 17, 2014.

In the months leading up to UN Climate Change Conference in Paris (COP 21), relevant special rapporteurs significantly stepped up their advocacy efforts. On the occasion of World Environment Day on June 5, 2015, twenty-seven mandate-holders issued a joint statement in which they called on States "to maintain language in the 2015 climate agreement that provides that the parties shall, in all climate change related actions, respect, protect, promote and fulfil human rights for all."[167] They also appealed on their moral obligations toward future generations:

> Climate change is one of the greatest human rights challenges of our generation, and it is our generation that must meet it. Indeed, the heads of governments and their climate negotiators represent the very last generation that can prevent catastrophic environmental harm to a vast array of human rights. We will continue to support them and all those working to protect human rights from this grave threat.[168]

A strategic move – demonstrating the manner in which various mechanisms can work in a mutually reinforcing manner allowing maximum cross-fertilization of separate agendas – was the invitation by the Philippines, the then Chair of the Climate Vulnerable Forum,[169] that is, the same group of countries that are particularly vulnerable to the impact of climate change and which therefore spearheaded the efforts to linking human rights and climate change through the adoption of the Malé Declaration, to a number of special rapporteurs to prepare a report on the effects of climate change on the full enjoyment of human rights agenda. This report was then submitted to COP 21.[170]

Besides the Special Rapporteur on human rights and the environment, several rapporteurs, namely the Special Rapporteur on the rights of indigenous peoples, the Special Rapporteur on the right to food and the Independent Expert on human rights and international solidarity, also participated in the Paris Climate Change Conference and delivered their messages in person, urging that the Paris Agreement incorporates a human rights-based approach.[171] The Special Rapporteur on human rights and the environment made a strong appeal, noting that "[e]very State in the climate negotiations belongs to at least one human rights treaty, and they must ensure that all of their actions comply with their human rights obligations. That

167 Joint statement by UN Special Procedures on the occasion of World Environment Day, June 5, 2015, Climate Change and Human Rights.

168 *Ibid.*

169 For further information on the Climate Vulnerable Forum, see www.thecvf.org (accessed May 7, 2019).

170 The United Nations Special Rapporteurs on the rights of persons with disabilities, C. Devandas Aguilar; the issue of human rights obligations relating to the enjoyment of a safe, clean, healthy, and sustainable environment, J. H. Knox; extreme poverty and human rights, P. Alston; the human right to safe drinking water and sanitation, L. Heller; and the Independent Expert on human rights and international solidarity, V. Dandan, The Effects of Climate Change on the Full Enjoyment of Human Rights, April 30, 2015.

171 A/HRC/31/52, para. 17.

includes their actions relating to climate change."[172] The mandate holder called on States "to honour their human rights obligations as they negotiate the climate agreement," and stressed that "including a reference to human rights in the agreement itself would be of great symbolic and practical importance."[173]

## 2 *The Paris Agreement: Recognizing Human Rights Obligations*

Around the same time that the link between climate change and displacement earned progressive recognition in climate agreements, references to human rights found their way into such agreements, drawing on the language of Human Rights Council resolutions, with some of the States of the sponsors' core group successfully pursuing this two-pronged approach, both in the Human Rights Council and the meetings of the COP under the UNFCCC.

As early as 2009, when the Human Rights Council just started consideration of climate change and human rights and the then High Commissioner for Human Rights, Navanethem Pillay, had presented her first report on the subject, the Maldives and Switzerland – to name those that are also spearheading the Council's initiatives – tried to infuse language on the relation between climate change and human rights, which had earlier been negotiated and agreed to in the Council within the climate change negotiation processes. The COP 15 draft negotiating text[174] included several human rights references, but the conference failed to reach an agreement on a new binding instrument. The draft negotiating text, which was revised further to the COP 15 deliberations, formed the basis for ensuing consultations for a new climate agreement. It included a reference to Council resolution 10/4 in its preambular part: "Noting resolution 10/4 of the United Nations Human Rights Council on human rights and climate change, which recognizes that human beings are at the centre of concerns for sustainable development, and the importance of respecting Mother Earth, its ecosystems and all its natural beings."[175] In addition, what was initially an operative paragraph,[176] moved in a later iteration of the negotiating text to the preambular part: "Mindful that the adverse effects of climate change have a range of direct and indirect implications for the full enjoyment of human rights, including living well, and that the effects of climate change will be felt

172 COP 21: "States' human rights obligations encompass climate change" – UN expert, Paris, December 3, 2015.

173 *Ibid.*

174 See inter alia FCCC/AWGLCA/2009/INF.1 of June 22, 2009, FCCC/AWGLCA/2009/INF.2, September 15, 2009.

175 FCCC/AWGLCA/2009/17, Annex I.

176 "These adverse effects [also] [further] [undermine the equitable development needs of present and future generations] [demand a more equitable utilization of the global atmospheric resource to reflect the needs of present and future generations], and have a range of direct and indirect implications for the full and effective enjoyment of human rights including the right to self determination, statehood, life, food and health and the right of a people not to be deprived if its own means of subsistence, particularly in developing countries." See FCCC/AWGLCA/2009/INF.2, September 15, 2009.

most acutely by those parts of the population that are already vulnerable owing to youth, gender, age or disability."[177] In a subsequent version of the negotiating text, both paragraphs were eventually merged to read:

> Noting resolution 10/4 of the United Nations Human Rights Council on "Human rights and climate change," which recognizes that the adverse effects of climate change have a range of direct and indirect implications for the effective enjoyment of human rights and that the effects of climate change will be felt most acutely by those segments of the population that are already vulnerable owing to geography, gender, age, indigenous or minority status and disability.[178]

This precise formulation was ultimately incorporated in the Cancún Agreements, which, while only a decision as opposed to a new binding instrument, were nevertheless an important advancement from a human rights perspective, even though the reference to Council resolution 10/4 contained in the preambular part was only "noted," which does not have any legal bearing. The Cancún Agreements, however, also emphasized in operative paragraph 8 that State parties in all climate-related actions should "fully respect human rights."[179] The inclusion of this obligation in the shared vision for long-term cooperative action of the Cancún Agreements, which forms the basis for all actions under the agreement on adaptation and mitigation as well as finance and technology, seeks to ensure that human rights considerations permeate States' considerations in their response to climate change.

The Cancún Agreements have been a significant step as it was the first time that human rights were explicitly included in the climate change regime. Even though some stronger elements that had figured in the draft negotiating text in Bonn could not be adopted,[180] this has to be contextualized, as some States opposed any reference to human rights on similar grounds they oppose reference to climate change in Human Rights Council resolutions.

The Paris Agreement, adopted on December 12, 2015, finally brought the two parallel tracks together as it recognized not only the linkages between climate change and displacement, but also called upon State parties to "respect, promote and consider their respective obligations on human rights" when addressing the challenges of climate change.[181] The recognition of the impact of climate change on human rights clearly sets out the obligations of States to those affected by climate change, calling for a human rights-based approach, which will ensure that any measure taken with regard to mitigation or adaptation does not interfere with the enjoyment of human rights by those concerned.[182]

177 FCCC/AWGLCA/2009/17, Annex I.
178 FCCC/AWGLCA/2010/6, May 17, 2010, Annex I.
179 FCCC/CP/2010/7/Add.1, March 15, 2011.
180 FCCC/AWGLCA/2010/14, August 13, 2010.
181 FCCC/CP/2015/10/Add.1, Annex, Paris Agreement.
182 S. McInerney-Lankford, Human Rights and Climate Change: Reflections on International Legal Issues and Potential Policy Relevance, 2013, p. 204 *et seq.*

## V INTERMEDIATE REMARKS

While these developments seem to demonstrate a growing international recognition of the effects of climate change on the enjoyment of human rights, some continue to invoke practical constraints to a human rights-based approach to climate change such as path-dependency,[183] while others seem concerned at importing politicization into what is considered a merely scientific debate through the human rights angle.[184]

Over time, nevertheless, it has to be said that the clear-cut separation between the human rights and the climate change frameworks that previously existed has now blurred somewhat and the seaming divide narrowed considerably. The cross-referencing in Council resolutions and the Paris Agreement attest to that effect. States such as Bangladesh and the Philippines, who are the main driving forces behind the Council's action on climate change, alongside the Maldives, which tabled the very first resolution, and Vietnam who joined the group of sponsors later on, persistently push for the recognition of the human rights impact of climate change, while others in the climate negotiation advocated for the inclusion of human rights considerations in the climate change framework.[185] A critical lesson from this process is that successful climate policy not only requires political will and substantive agreement but also the right timing and sequencing of events.

The significance of the ultimate recognition of the climate change–displacement nexus on the one hand and the climate change–human rights interrelation on the other is that this makes it finally possible (and imperative) to design policies and programmes that place the human being at the centre when addressing the effects of climate change; that is, it was the precondition toward adopting a human rights-based approach to climate displacement. The inclusion of the human dimension (i.e. the recognition that people will increasingly be on the move as a result of the adverse effects and that their rights have to be upheld) significantly shapes the responses to climate change and the resulting displacement. This tenet has been the starting point of the Peninsula Principles.

183 National Office of Oceanic and Atmospheric Research, Global Warming and Hurricanes, www.oar.noaa.gov/spotlite/archive/spot_gfdl.html (accessed June 2, 2019).

184 M. Limon, Human Rights and Climate Change: Constructing a Case for Political Action, 2009, p. 459 *et subseq.*

185 S. Duycks, UN Human Rights Council declares climate a priority, *Climate Home News*, June 13, 2016.

# 3

# Protection Challenges and Policy Options

## I NORMATIVE FRAMEWORK AND PROTECTION CHALLENGES

Climate displacement, whether international or domestic, raises a number of protection challenges at various levels. Some of them may be addressed through the existing international frameworks, including the international refugee regime or human rights law as well as nonbinding normative or policy frameworks, while, in a number of instances, protection gaps may arise.

The set of challenges is different, subject to the type of displacement. Climate displaced persons crossing international borders may or may not fall within the classical refugee law framework depending on the specific circumstances of their flight. This may be the case when the legal requirements under the 1951 Refugee Convention, as modified by the 1967 Protocol,[1] are met, for example in the case of refugee flows provoked by climate-induced armed conflict. In such instances, international or regional refugee instruments are applicable.[2] As former High Commissioner for Refugees António Guterres stated, the causes for displacement are often

> interlinked, overlapping and mutually reinforcing factors that accumulate and contribute to eventually driving people from their homes. Violent conflicts, just like human rights violations and persecution, do not erupt in a vacuum. They are the final result of a complex interaction of problems ranging from inequality and marginalization, lack of good governance and rule of law, competition over shrinking resources in a context of population growth and chaotic urbanization – often exacerbated by environmental degradation and the effects of climate change.[3]

1 For the full text of the 1951 Refugee Convention, see www.unhcr.org/protect/PROTECTION/3b66c2aa10.pdf (accessed April 18, 2019).

2 UNHCR, Climate Change, Natural Disasters and Human Displacement: A UNHCR Perspective, 2009, p. 2 *et subseq*.

3 A. Guterres, Opening Remarks, High Commissioner's Dialogue on Protection Challenges: Understanding and Addressing Root Causes of Displacement, Geneva, December 16, 2015.

It will be much more difficult to invoke refugee status for an individual when the drivers of displacement cannot be associated with armed conflict. According to the UNHCR, the element of persecution required under the Refugee Convention encompasses the conscious withholding or obstruction of assistance for a victim of a natural disaster with a view to punishing them for reasons of race, religion, nationality, membership of a particular social group or political opinion.[4] Very few cases of cross-border climate displacement may actually fall in this particular category, even though climate change and environmental degradation may constitute a humanitarian circumstance that would require protection.[5]

Regional instruments (i.e. the 1969 Organisation of African Unity Convention Governing the Specific Aspects of Refugee Problems in Africa and the 1984 Cartagena Declaration on Refugees) provide a nuanced refugee definition also covering events seriously disturbing public order. The latter furthermore also includes massive human rights violations and internal conflicts and may therefore offer protection in some cross-border climate displacement scenarios.

Protection challenges also exist in relation to populations that could be forced to relocate as their State's territory ceases to exist owing to rising sea levels, as may be the case for some SIDS. If the State would disappear as a result of the loss of its territory, its population would de facto become stateless in accordance with Article 1 of the 1954 Convention Relating to the Status of Stateless Persons, which defines a stateless person as someone who is not considered a national by any state under the operation of its law. Apart from the legal perspective, such situations, which would be unprecedented, would also raise issues of preservation of cultural heritage and cultural identity.[6]

Complementary forms of protection are alternative forms of protection for those climate displaced persons who do not formally qualify as refugees but who cannot return or whose return is not advisable. Some States have adopted a pragmatic approach to arising international protection needs by devising administrative or legal mechanisms that would allow to regularize the stay of persons concerned.[7] While these individual responses by some States have to be welcomed, they can lead to a proliferation of statuses that are not recognized within the international nomenclature and the need for a coordinated approach remains in order to ensure the same degree of protection for all climate displaced persons irrespective of their place of destination.

4 UNHCR, Forced Displacement in the Context of Climate Change: Challenges for States under International Law, 2009, p. 9 *et subseq*.

5 B. L. Burson, [2014] NZIPT 501370–371, Immigration and Protection Tribunal, Auckland, New Zealand, para. 27 *et seq*.; see also B. L. Burson, [2014] NZIPT 800517–520, Immigration and Protection Tribunal, Auckland, New Zealand; B. L. Burson, [2013] NZIPT 800413, Immigration and Protection Tribunal, Auckland, New Zealand.

6 UNHCR, Climate Change and Statelessness: An Overview, 2009.

7 UNHCR, EC/50/SC/CRP.18, Executive Committee of June 9, 2000, Complementary Forms of Protection: Their Nature and Relationship to the International Refugee Protection Regime.

Climate displaced persons who do not cross internationally recognized borders are referred to as IDPs. They fall within the jurisdiction of the State concerned and are therefore better protected than those displaced who cross international borders, as they enjoy protection under applicable domestic laws as well as international human rights law. This includes the human rights enshrined in the Universal Declaration of Human Rights as well as in international human rights treaty law, notably the two international covenants (i.e. the International Covenant on Civil and Political Rights (ICCPR) and the International Covenant on Economic, Social and Cultural Rights (ICESCR)). Article 4(2) of the ICCPR stipulates that a number of the rights laid down in the Convention are non-derogable even in times of public emergency that threatens the life of the nation. These include the right to life; freedom from torture or cruel, inhuman and degrading treatment or punishment; freedom from slavery or to be held in servitude; recognition as a person before the law; and nondiscriminatory treatment. Derogation of other rights is also strictly limited to situations of public emergency that threaten the life of the nation. The conditions for such derogations in the context of a state of emergency are further detailed in the Human Rights Committee's General Comment 29 on that matter.[8] The ICESCR, while following a different approach (i.e. it does not stipulated non-derogable rights), sets out minimum core obligations of States to ensure the satisfaction of minimum levels of each right: for instance, minimum essential food that is sufficient, nutritionally adequate and safe to ensure freedom from hunger,[9] essential basic shelter and housing, including sanitation,[10] and the right not to be arbitrarily evicted from one's house.[11]

Internal displacement as a result of a situation where the effects of climate change are exacerbated by armed conflict would moreover be covered by international humanitarian law, in particular the Geneva Conventions of 1949 and their Additional Protocols of 1977. Common Article 3 of the Geneva Conventions covers situations of non-international armed conflicts, whereas the Geneva Convention relative to the Protection of Civilian Persons in Time of War (Fourth Geneva Convention) is of specific relevance to the protection of IDPs in international armed conflicts.[12]

Internal climate displacement nevertheless raises a number of practical and normative challenges as the return of the displaced may not be possible and as climate displacement may disproportionately affect the most vulnerable in a country.

[8] CCPR/C/21/Rev.1/Add.11.

[9] E/1991/23 of December 14, 1990; E/C.12/1999/5 of May 12, 1999.

[10] E/1991/23 of December 14, 1990; E/1992/23 of December 13, 1991.

[11] E/1998/22 of May 20, 1997; UN-Habitat/UNOHCHR, Forced Evictions, Fact Sheet No. 25, Rev.1, 2014; E/CN.4/Sub.2/2005/17/Add.1, para. 16; E/1998/22; COHRE, Forced Evictions: Violations of Human Rights No. 10, 2006; COHRE, Forced Evictions: Violations of Human Rights No. 7, 1994.

[12] Geneva Convention Relative to the Protection of Civilian Persons in Time of War, adopted on August 12, 1949, by the Diplomatic Conference for the Establishment of International Conventions for the Protection of Victims of War, held in Geneva from April 21 to August 12, 1949.

Displacement that becomes permanent in nature requires specific solutions for which States may be ill-prepared and lack the required policy and legal frameworks.[13] One of the key practical (and legal) challenges in connection with permanent displacement are housing, land and property rights, and issues of livelihood of the displaced, which are relevant in all displacement settings, but even more so when return is not possible because of irreversible effects of climate change and the ensuing need to find durable solutions. In particular, even in cases of permanent displacement, climate displaced persons remain entitled to their rights related to housing and property restitution/compensation, which, as emphasized by the Special Rapporteur on the human rights of internally displaced persons, is "so vital to durable solutions."[14] The Special Rapporteur's findings illustrate the broader challenges faced by States, both individual and as a result of their collective responsibility, which "includes addressing food crisis brought on or aggravated by loss of arable land, droughts, floods, and loss of biodiversity, but it also entails putting in place systems of social protection for persons who have become landless, homeless, deprived of livelihoods – and more often than not also internally displaced."[15] Another key question, in the context of relocations of permanently displaced persons due to the effects of climate change, relates to the human rights of the host communities. In order to address these challenges resulting from climate displacement in an adequate manner, there is a pressing need to develop national and international normative, institutional and implementation frameworks. Such frameworks need to be people-centered and follow a human rights-based approach.

The protection challenges have led to some unique approaches, which should in fact alarm the international community. The low-lying Maldives, for instance, have started acquiring land abroad to relocate their nationals in light of the anticipated climate change related sea-level rise.[16] The issue of relocation certainly needs to be considered in relation to those for whom return will become impossible, as in the case of submerged islands or permanently flooded territories.[17] Such a course of action is however questionable in several of its aspects, and while it may eventually constitute a precedent for wealthy States threatened in their existence, which will try to reestablish themselves elsewhere, the question is how others who lack the necessary funds and resources could handle such situations. Relocation in internal

13 K. Koser, Gaps in IDP Protection, 2008, p. 17; R. Zetter, Legal and Normative Frameworks, 2008, p. 62 *et subseq.*

14 A/HRC/16/43, para. 61.

15 *Ibid.*, para. 65.

16 J. Evans, Climate change drives Maldives to buy land, Financial Times, November 10, 2008; IPCC, Climate Change 2007: Impacts, Adaptation and Vulnerability, 2007, p. 672; in relation to the Tuvalu case, see S. Leckie, Climate-Related Disasters and Displacement: Homes for Homes, Lands for Lands 2009, p. 9 *et subseq.*

17 The Australian Greens party started an initiative and failed to grant special visas to Tuvalan climate change displaced; J. Evans, Climate change drives Maldives to buy land, Financial Times, November 10, 2008; see also S. Leckie, Two Climate Change and Displacement Scenarios: The Choice Is Ours, A Presentation to the Turning Point, 2009; see also, M. Loughry, J. McAdam, Kiribati: Relocation and Adaptation, 2008, p. 51 *et subseq.*

displacement also requires concrete guidelines and protocols to ensure that the rights of the displaced and the host communities are equally protected.[18]

## II EXISTING FRAMEWORKS AND POLICY OPTIONS

### 1 *Guiding Principles on Internal Displacement*

The Guiding Principles on Internal Displacement,[19] adopted in 1998, provide a normative framework to address protection challenges in the context of internal displacement. While nonbinding, the Guiding Principles are now widely recognized as "an important international framework for the protection of internally displaced persons."[20] They embody international law and thereby constitute an authoritative framework for the protection of the rights of internally displaced persons. They explicitly refer to situations of "natural and human-made disasters" and are therefore applicable to climate displacement.[21] The Principles span all phases of internal displacement from prevention (section II – principles relating to protection from displacement), to the actual stage of displacement (section III – principles relating to protection during displacement) and the required assistance (section IV – principles relating to humanitarian assistance) to the post-displacement phase (section V – principles relating to return, resettlement and reintegration).[22]

Climate displacement, for as long as it remains internal, could thus simply be seen as a subcategory of internal displacement. However, some questions remain regarding the coerced nature of displacement as a result of climate change, particularly if it takes place in the early stages of environmental degradation as an adaptation strategy.[23] Another issue relates to the question of arbitrariness of the displacement. The Guiding Principles explicitly restate the right not to be arbitrarily displaced: "Every human being shall have the right to be protected against being arbitrarily displaced from his or her home or place of habitual residence." They moreover provide that "[a]ll authorities and international actors shall respect and ensure respect for their obligations under international law, including human rights

[18] For examples of climate-related relocations, see, e.g., M. Ezra, G.-E. Kiros, Rural Out-Migration in the Drought Prone Areas of Ethiopia: A Multilevel Analysis, 2011; G. P. Marchildon *et al.*, Drought and Institutional Adaptation in the Great Plains of Alberta and Saskatchewan, 1914–1939, 2008; S. F. Martin, Managing Environmentally Induced Migration, 2009, p. 369 *et seq.*, UNEP, Livelihood Security, Climate Change, Migration and Conflict in the Sahel, 2001, p. 71 *et seq.*, see also Brookings Institution, Internal Displacement and the Construction of Peace, Summary Report, Bogota, Colombia, 2008.

[19] E/CN.4/1998/53/Add.2.

[20] A/RES/60/1, para. 132; see also A/RES/64/162, preamble and para. 10.

[21] E/CN.4/1998/53/Add.2, para. 2.

[22] E/CN.4/1998/53/Add.2.

[23] R. Black, *et al.*, Climate Change: Migration as Adaptation, 2011, p. 447 *et seq.*; J. Barnett, M. Webber, Accommodating Migration to Promote Adaptation to Climate Change, 2010.

and humanitarian law, in all circumstances, so as to prevent and avoid conditions that might lead to displacement of persons." Generally, addressing internal displacement as envisaged by the Guiding Principles seeks to prevent certain actions (i.e. human rights violations), including those resulting from large-scale development projects, in order to avoid situations of displacement. Their focus therefore is on the arbitrary nature of displacement. With regard to climate displacement, prevention instead requires positive actions, and there is therefore a need to outline specifically the obligations of States when it comes to taking preventative measures and to mitigate climate displacement.[24] This is a fundamental difference, not necessarily unbridgeable from a normative point of view, but one that elicits policy and operational gaps that need to be addressed.[25] One merit however of the Guiding Principles, and which also points to possible analogies for addressing any potential gaps, is that they effectively bundle existing yet unspecific protection norms to cover situations of internal displacement and they are therefore considered "a model for the process of aggregating and adapting the norms and principles from a wide range of international instruments to protect the rights of the 'environmentally displaced'."[26]

## 2 *The Pinheiro Principles*

In 2005, the United Nations Sub-Commission endorsed the so-called Pinheiro Principles,[27] the United Nations Principles on Housing and Property Restitution for Refugees and Displaced Persons, which encapsulate and restate the existing normative framework of international human rights and humanitarian law, and relevant international legislation and practice in relation to a specific set of rights of displaced persons (i.e. housing, land and property rights) that are considered a prerequisite for their return. While not specifically drafted to address climate displacement, the Pinheiro Principles can nevertheless be drawn upon for guidance on how to best ensure the international and national protection of housing, land and property rights in post-disaster settings induced through climate change.[28] As a matter of fact, the Pinheiro Principles have informed the drafting of the Peninsula Principles on Climate Displacement within States, as specifically

24 *Cf.* B. G. Ramcharan, Preventive Human Rights Strategies, p. 27 *et seq.*

25 See also, G. Hugo, Migration, Development and Environment, 2008, p. 49.

26 R. Zetter, Legal and Normative Frameworks, 2008.

27 E/CN.4/Sub.2/2005/17.

28 Existing international human rights standards relevant for the bundle of HLP rights comprise the UDHR, the ICCPR and the ICESCR as well as important interpretive standards such as CESCR General Comment No. 4 on the Right to Adequate Housing, CESCR General Comment No. 7 on Forced Evictions and CESCR General Comment No. 15 on the Right to Water; see also S. Leckie, Climate-Related Disasters and Displacement: Homes for Homes, Lands for Lands, 2009, p. 3; FAO/IDMC/OCHA/OHCHR/UN-Habitat/UNHCR, Handbook, Housing and Property Restitution for Refugees and Displaced Persons, Implementing the "Pinheiro Principles," 2007, p. 16 *et subseq.*; K. Hassine, Regularizing Property Rights in Kosovo and Elsewhere, 2010.

indicated in the preamble, which attests to the former's particular relevance for situations of climate displacement.[29]

Drafting of the Pinheiro Principles originated in 1997 when the United Nations Committee on the Elimination of Racial Discrimination (CERD) proposed to the Sub-Commission on the Promotion and Protection of Human Rights (the body that subsequently endorsed the Pinheiro Principles) to study "the return of refugees' or displaced persons' property."[30] Following this proposal, a year later, the Sub-Commission adopted resolution 1998/26 on housing and property restitution in the context of the return of rights for refugees and internally displaced persons and thereby confirmed its intention to further study the question of restitution. In 1999, the Commission on Human Rights (the Sub-Commission's parent body) encouraged the Sub-Commission to continue its work on housing and property restitution in the context of the return of refugees and IDPs. It was only in 2001 however that the Sub-Commission mandated a first working paper on the topic. Paulo Sérgio Pinheiro was entrusted in the subsequent year with the elaboration of draft principles or guidelines on housing and property restitution and, subsequently, presented his final text in 2005.

The Pinheiro Principles seek to provide practical guidance on how best to address the complex legal and technical issues surrounding housing, land and property restitution. They are also applicable in situations of natural disasters or in relation to climate displacement. This set of principles is a key instrument for an efficient rights-based and people-oriented displacement management, a fortiori since they are based on and draw from existing and applicable international human rights norms and humanitarian law, just like the Guiding Principles on Internal Displacement.[31] Their scope is not limited to internal displacement, but also includes forced cross-border mobility.

Rising sea levels and storms, for instance, are direct causes of the flooding of large parts of territories, which may at times be of permanent nature, the salination of freshwater resources and diminishing habitable and/or cultivable land. Particularly vulnerable are coastal settlements, low-lying island States or settlements in low-lying megadeltas.[32] Such degrading environmental conditions are often causal for population movements and, at the same time, directly impact housing, land and property rights.[33] The protection of communities that are relocated away from hazardous

[29] The Peninsula Principles, http://displacementsolutions.org/peninsula-principles (February 5, 2019).

[30] E/CN.4/Sub.2/1997/31, May 27, 1997, annex, p. 4; for an overview of the discussions in the Committee, see CERD/C/SR.1189 of March 8, 1997.

[31] FAO/IDMC/OCHA/OHCHR/UN-Habitat/UNHCR, Handbook, Housing and Property Restitution for Refugees and Displaced Persons, 2007, p. 12 *et subseq.*; K. Hassine, S. Leckie, The United Nations Principles on Housing and Property Restitution for Refugees and Displaced Persons: A Commentary, p. 8 *et subseq.*, as well as the *travaux préparatoires* part 2, p. 173 *et subseq.*

[32] IPCC, Climate Change 2007: Impacts, Adaptation and Vulnerability, 2007, p. 333.

[33] According to UN-Habitat a total of 211 million people affected by natural disasters in 1991–2000 were in developing countries. See UN-Habitat, Global Report of Human Settlements: Enhancing Urban Safety and Security, 2007.

zones needs to be guaranteed, in accordance with the Pinheiro Principles. This includes protection against forced evictions carried out without appropriate forms of legal or other protection as well as the need for adequate consultation with affected persons.

Of particular relevance in this connection is Principle 10 of the Pinheiro Principles, which affirms the right of all displaced, including refugees, to return in safety and dignity to their former homes and houses.[34] In the period following the 2004 tsunami in the Indian Ocean for instance, some governments misused adaptation strategies.[35] With the argument of an imminent security threat, authorities adopted policies that established coastal buffer zones, by which they prevented low-income affected persons returning to their areas of origin and forced communities in coastal areas to permanently relocate, thereby disrupting their livelihoods and generating and exacerbating social tensions.[36] By way of illustration, in Sri Lanka, the majority of the half million people affected by the tsunami lived within the 100 m buffer zone established by the Government. Fishing communities were particularly affected by these measures and were permanently prevented from returning to their original homes. Moreover, such buffer zones were often transferred to higher-income residential, commercial or industrial use, as for instance the tourism industry expanded its operations into the newly vacated land. Unjustifiable restrictions to the right to return may furthermore amount to forced evictions,[37] against which any person is protected by international law.[38] This incompatibility is implicitly recognized in Principle 8 of the Pinheiro Principles, which stipulates the right to adequate housing, as well as in Principle 6, the right to privacy and respect for the home, and is more explicitly stated in Principle 5, which provides for the protection from displacement and places upon governments the obligation to take necessary steps in this respect. Abandonment laws, which result in the seizure of vacated property after certain periods of time, are another example of incompatible measures observed in post-disaster situations. Moreover, reconstruction and rebuilding efforts are often slow, also because of a lack of availability of material, tools and expertise for repair and reconstruction, and may impair the right to return. In fact, temporary relocation should only last for as long as it is absolutely necessary.[39]

34 See also E/CN.4/Sub.2/2005/17/Add.1, para. 38 *et subseq.*

35 S. Leckie, An Integrated Approach to Climate Change Demands That Human Rights and Adaptation Strategies Are Pursued Hand-in-Hand, 2008, p. 18; see also A. Naik, E. Stigter, F. Laczko, Migration, Development and Natural Disasters: Insights from the Indian Ocean Tsunami, 2007; Displacement Solutions, Regulatory Obstacles to Rapid and Equitable Emergency and Interim Shelter Solutions after Natural Disasters, 2011, p. 29 *et subseq.*

36 S. Leckie, Human Rights Implications, 2008, p. 18.

37 *Forced eviction* is the permanent or temporary removal against their will of individuals, families and/or communities from the homes and/or land which they occupy, without the provision of, and access to, appropriate forms of legal or other protection, see E/CN.4/Sub.2/2005/17/Add.1, para. 16.

38 See also IRIN News, Global: Taking on the land-grabbers, October 26, 2010.

39 A/64/255, para. 55; S. Leckie, Returning Home: Housing and Property Restitution Rights of Refugees and Displaced Persons, 2003, p. 45 *et seq.*

Another provision that often comes into play in those circumstances where governments allow for the return to the places and homes of origin is the right to adequate housing reiterated in Principle 8 of the Pinheiro Principles. In line with the obligation to adopt positive measures aimed at alleviating the situation of the displaced, States have to provide safe and adequate shelter to victims to temporarily bridge the period before actual return is possible. Such temporary housing arrangements need to be in conformity with international standards, in particular with regard to access to water, sanitation and other basic services. The right to adequate housing is of relevance not only *during* displacement, but in the course of the entire displacement cycle. In this connection, the UNHCR for example advocates for the temporary allocation of land to displaced individuals or communities during their displacement, with a view to encouraging and facilitating self-sustainability.[40]

Reconstruction and rehabilitation efforts need to be carried out in respect of the overarching principle of nondiscrimination[41], as affirmed in Principle 3 of the Pinheiro Principles, as a lack of home or residence at times prevents people's access to credit and provision of financial assistance. At the same time, return itself must be allowed without discrimination. This nondiscrimination provision must also be read in conjunction with the subsequent Principle 4 stipulating the right to equality between men and women. Natural disasters disproportionally affect women and children as well as other vulnerable groups, such as older persons.[42] Discriminatory and/or harmful traditional norms or practices such as discriminatory inheritance laws often prevent women survivors of disasters acquiring their legitimate property rights and are often disregarded in property restitution processes.[43]

Other obligations of governments during displacement result from the antidestruction and protection clauses contained in Principles 5.3 and 5.4 of the Pinheiro Principles in conjunction with Principles 17 and 21. Authorities need to take

40 UNHCR Executive Committee, Conclusion on Legal Safety Issues in the Context of Voluntary Repatriation of Refugees, October 8, 2004.

41 The right to nondiscrimination is protected in many major international human rights instruments, incl. Art. 2 of the UDHR, Art. 2 of the ICESCR and the ICCPR and is also prohibited under the International Convention on the Elimination of All Forms of Racial Discrimination (ICERD), CEDAW, Art. 2 of the Convention on the Rights of the Child (CRC), Art. II of the American Declaration on the Rights and Duties of Man, Art. 1 of the American Convention on Human Rights, Art. 2 of the African Charter on Human and Peoples' Rights and Art. 14 of the European Convention for the Protection of Human Rights and Fundamental Freedoms.

42 See Displacement Solutions/ARA-JPD/OHCHR, Side Event on Building Climate Resilience, The Rights of Groups in Focus, March 18, 2015; PR/OHCHR/20March2015, New framework for Disaster Risk Reduction; A/HRC/30/43/Add.3, para. 54 *et subseq.* and 114; A/HRC/33/44/Add. 1, para. 97; Brookings Institution, The Year That Shook the Rich: A Review of Natural Disasters in 2011, March 2012, p. 115 *et seq.*

43 The Commission on Human Rights in its resolution 2005/25, Women's Equal Ownership, Access to and Control over Land and the Equal Rights to Own Property and to Adequate Housing, affirmed "that discrimination in law and practice against women with respect to having access to, acquiring and securing land, property and housing, as well as financing for land, property and housing, constitutes a violation of women's human right to protection against discrimination and may affect the realization of other human rights," E/CN.4/RES/2005/25.

adequate measures to protect the property rights of those displaced by natural disasters and to preserve houses and properties from destruction, looting and secondary occupation. Such measures may also include a temporary freeze on the transfer of land and housing. The post-tsunami setting has been a good illustration of how such disruptive circumstances are used as pretext not only for forced and arbitrary evictions, but also for massive land grabs.[44]

Upon return, authorities need to ensure security of tenure in line with Principle 15 of the Pinheiro Principles, which includes the establishment of national multipurpose cadastres or other appropriate systems for the registration of housing, land and property rights. In case secondary occupation by other displaced persons has occurred or disputes over properties arise in the process of return, the Pinheiro Principles, with their detailed provisions on the establishment of restitution mechanisms and the conduct of the restitution process, provide the required framework and guidance to law- and policymakers to fulfill their obligations to provide an effective remedy for arbitrary deprivation of property in the form of restitution and/or compensation. Disputes arise not only from secondary occupation and the acquisition of housing, land and property by third parties acting in good faith, but also in light of landmarks that were destroyed or removed during or in the aftermath of the disaster, and the unavailability of titling systems as well as the lack of property records. In such situations of mass displacement, where little or no documentary evidence may be available, the Pinheiro Principles provide that the authorities may independently establish the facts related to the undocumented restitution claims.[45]

Resolving large numbers of claims in a timely, prompt, adequate and fair manner, without endangering the livelihood of the concerned claimants, challenges and overstrains every judicial system, and even more so in the developing world. Most judicial systems are not designed to deal with mass claims. States, sometimes in conjunction with or the support of the international community, as in the post-conflict property regularization in Kosovo,[46] are therefore required to provide for precautionary measures,[47] and to establish competent, judicial and administrative

44 S. Leckie, The Great Land Theft, 2005, p. 15 *et seq.*; S. Leckie, Returning Home: Housing and Property Restitution Rights of Refugees and Displaced Persons, 2003, p. 47 *et seq.*

45 K. Hassine, S. Leckie, The United Nations Principles on Housing and Property Restitution for Refugees and Displaced Persons: A Commentary, 2016, p. 62 *et subseq.* and p. 129; S. Leckie, Returning Home: Housing and Property Restitution Rights of Refugees and Displaced Persons, 2003, p. 47 *et seq.*

46 K. Hassine, Regularizing Property Rights in Kosovo and Elsewhere, 2010; K. Hassine, Housing and Property Directorate/Claims Commission in Kosovo, 2009.

47 ECHR judgment of November 30, 2004, *Öneryıldız* v. *Turkey*; Appl. No. 48939/99, para. 89 *et subseq.*; ECHR judgment of March 20, 2008, *Budayeva and Others* v. *Russia*, Appl. No. 15339/02, para. 131: "The obligation on the part of the State to safeguard the lives of those within its jurisdiction has been interpreted so as to include both substantive and procedural aspects, notably a positive obligation to take regulatory measures and to adequately inform the public about any life-threatening emergency, and to ensure that any occasion of the deaths caused thereby would be followed by a judicial enquiry"; ECJ judgment of July 25, 2008, *Dieter Janecek* v. *Freistaat Bayern*, Case C-237/07; BVerwG 7 C 36.07 (2007).

procedures for considering restitution claims. The Pinheiro Principles contain provisions that facilitate the design and establishment of adequate claims settlement mechanisms, including for those instances where massive numbers of claims will have to be managed (Principle 12).

While the application of the Pinheiro Principles provides for protection in favor of all displaced persons irrespective of the causes of displacement and their location (i.e. including as a result of changing climatic conditions), they are not specifically designed to address situations of climate displacement and focus particularly on housing, land and property rights. The Guiding Principles by contrast address the broader issue of internal displacement, but do not, beyond the general principle, provide practical guidance in terms of the handling of climate displacement situations.[48]

## 3 *From the Global Guiding Framework to the Nansen Initiative*

### a Global Guiding Framework

The impact of climate change on displacement had long been a concern for the UNHCR.[49] António Guterres, then UN High Commissioner for Refugees and later UN Secretary-General, in 2007 at the occasion of the 3rd Symposium on Corporate Social Responsibility and Humanitarian Assistance in Tokyo underscored that there were three reasons why people were on the move: first, poverty, which fuels irregular migration and is providing fertile ground for people smugglers and human traffickers; second, climate change and environmental degradation, while noting that there was no political will to address this matter; and third, conflict and persecution, which – when linked to other causes (e.g. water scarcity, as in Darfur)[50] – can have devastating effects.[51]

On the second main point, the High Commissioner further elaborated that

> [w]hen we consider the different models for the impact of climate change, the picture is very worrying. The need for people to move will keep on growing. One need only look at East Africa and the Sahel region. All predictions are that desertification will expand steadily. For the population, this means decreasing livelihood prospects and increased migration. All of this is happening in the absence of international capacity and political will to respond. Discussions about climate change are taking place, but there is still no real strategy for how to cope with it.[52]

48 E/CN.4/1998/53/Add.2.

49 N. Hall, Displacement, Development, and Climate Change, 2016, p. 57 *et seq.*

50 B. Bromwich, Environmental Degradation and Conflict in Darfur: Implications for Peace and Recovery, 2008, p. 22 *et seq.*; H. Welzer, Klimakrieg, Wofür im 21. Jahrhundert getötet wird, 2010, p. 94 *et seq.*; see also, e.g., UNEP, Livelihood Security, Climate Change, Migration and Conflict in the Sahel, p. 71 *et seq.*; see also J. O. Simel, Le pastoralisme et les defis du changement climatique, 2009, p. 30 *et seq.*

51 A. Guterres, UN High Commissioner for Refugees, Keynote Speech at the Third Symposium on Corporate Social Responsibility and Humanitarian Assistance, Tokyo, November 26, 2007.

52 *Ibid.*

Similarly, Erika Feller, the then Assistant High Commissioner for International Protection, in June 2008 at the occasion of the 42nd Meeting of the Standing Committee, a body that examines thematic issues and reviews the UNHCR's activities and programs and, importantly, is in a position to adopt appropriate decisions and conclusions on matters that it deems of concern,[53] raised the issue of revalidating the protection objectives as well as the framework of law and principles. Her statement, already then, provided a snapshot of the quandary of the wide-ranging protection challenge of climate displacement for the years if not decades ahead:

> The challenge of protecting refugees continues to grow as conflicts proliferate, together with the factors that provoke them – from intolerance, bad governance and disrespect for human rights, to related issues like environmental degradation and increased competition for diminishing resources. We are warned increasingly of the indelible "footprint of climate change" and "looming water wars". New terminology is entering the displacement lexicon with some speed.

She then continued to address the notional discussion by saying that

> [t]he talk is now of "ecological refugees", "climate change refugees", the "natural disaster displaced". This is all a serious context for UNHCR's efforts to fulfill its mandate for its core beneficiaries, that is refugees fleeing persecution and violence or trying to return in precarious situations of peace, persons without an effective nationality, and IDP populations for whom we have particular responsibilities.[54]

On November 23, 2011, during a briefing to the UN Security Council on "Maintenance of International Peace and Security: 'New Challenges to International Peace and Security and Conflict Prevention',"[55] the then UN High Commissioner for Refugees seized the opportunity to alert that changing patterns of movement in conjunction with a growing scale of human displacement not only pose a problem to international peace and security, but require imperative action by the international community.

In this same briefing, he outlined that "there is little value in posing the simplistic question, 'how many people are going to be displaced by climate change?', rather linkages must be understood and acted upon like 'the potential for conflict within and between states by intensifying the level of competition for scarce resources, including water, grazing and arable land'." He also referred to the "so-called 'water

53 In October 1995, the Executive Committee established a Standing Committee (see A/AC.96/860) to replace the sub-committees on international protection and on administrative and financial matters. The chairing of the Standing Committee is shared by the Executive Committee chairperson and the vice-chairpersons. The Standing Committee is scheduled to meet three times a year. See also www.unhcr.org/standing-committee-meetings.html (accessed March 24, 2019).

54 *Ibid.*

55 Statement by A. Guterres, UN High Commissioner for Refugees, UN Security Council Briefing, Maintenance of International Peace and Security: New Challenges to International Peace and Security and Conflict Prevention, 2011.

wars' over transboundary freshwater reserves, with an evident potential to uproot large numbers of people. According to the UN Environment Programme, the conflict in Darfur has been at least partially driven by the process of climate change, environmental degradation and the struggle for access to land and water."[56] Concluding, the High Commissioner called to refocus attention and efforts on "manifestations of climate change [that] are interacting with and reinforcing other global imbalances, so as to produce some very powerful drivers of instability, conflict and displacement."[57]

These concerns laid the ground for the UNHCR's initiative to elaborate a Global Guiding Framework. In the lead-up to the 2011 Ministerial Conference, the UNHCR commissioned new research on climate change and displacement. A preparatory meeting was furthermore held in Bellagio (i.e. an expert Meeting on Climate Change and Displacement) to discuss gaps in the international response to the evolving phenomena.[58]

There was an expectation that the Ministerial Conference would generate some kind of support for institutional changes at the level of the UNHCR in order to allow the organization to specifically address displacement resulting from natural disasters, which is not within the scope of its initial mandate. In other words, the question of lead agency in providing assistance and protection to those affected, which UNHCR had assumed repeatedly but which was determined until then on a case-by-case basis in the IASC, was to be considered.[59]

In the conference's background note, the UNHCR had raised the key issues:

> Would it be useful for States, UNHCR and other relevant actors to develop a global guiding framework or instrument to apply to situations of displacement across borders other than those covered by the 1951 Convention? If so, should this be limited to displacement relating to climate change and natural disasters, or could it be broader? Could temporary or interim protection arrangements be useful? If so, in which situations?[60]

Yet, States took no action to empower the UNHCR to become the lead agency on displacement due to natural disasters and/or climate change, and only acknowledged that challenges in providing protection and achieving solutions continued to be serious, interconnected and complex.[61] The UNHCR was merely mentioned alongside other relevant stakeholders in the context of reinforcing cooperation.

56 *Ibid.*

57 *Ibid.*; see also, e.g., UNEP, Livelihood Security, Climate Change, Migration and Conflict in the Sahel, p. 71 *et seq.*

58 J. McAdam, Creating New Norms on Climate Change, Natural Disasters and Displacement: International Developments 2010–2013, 2014, p. 14.

59 N. Hall, Climate Change and Organizational Change in UNHCR, 2011, p. 110; N. Hall, Displacement, Development, and Climate Change, 2016, p. 62 *et seq.*

60 HCR/MINCOMMS/2011/08, Background Note for the Roundtables, p. 4.

61 UNHCR, Summary of Deliberations on Climate Change and Displacement, Bellagio, 2011.

Thus, in spite of the express need to deepen the understanding of evolving patterns of displacement and to agree upon ways to respond to the challenges faced, the Ministerial Conference did not lead to actual changes. One of the reasons why States were reluctant to confer upon UNHCR additional responsibilities and to expand its mandate, was that internal displacement, which would make up most of the climate displacement, remained a highly politically sensitive matter for reasons of potential interference into domestic affairs.[62]

### b The Nansen Initiative

The failure of the Global Guiding Framework is intimately linked to the launching of the Nansen Initiative in 2012. While the UNHCR was pushing for this new framework, Switzerland together with Norway convened the Nansen[63] Conference on Climate Change and Displacement in Oslo in June 2011 as a result of which the Nansen Principles were adopted. The Oslo conference was not *stricto sensu* a preparatory conference of the UNHCR's Ministerial Conference, unlike the Bellagio expert meeting, but its results fed into the Global Guiding Framework discussions, and while the UNHCR had no formal role in this conference, it was clearly connected to the UNHCR's agenda.

These Nansen Principles outline core approaches for addressing both internal and international displacement. Recognizing that the Guiding Principles on Internal Displacement already cover those displaced internally by natural disasters and other environmental disruptions, the Nansen Principles therefore urged action on cross-border movements: "A more coherent and consistent approach at the international level is needed to meet the protection needs of people displaced externally owing to sudden-onset disasters. States, working in conjunction with

[62] A/AC.96/1104, September 20, 2011, para. 33; EC/62/SC/CRP.19, 2011, para. 4; W. Kälin, From the Nansen Principles to the Nansen Initiative, 2012, p. 49.

[63] The conference was named after Fridtjof Nansen, who was appointed in 1921 by the League of Nations to become High Commissioner for Refugees and earned great merits in dealing with the Russian Refugee Crisis, in which a Soviet decree of 1921 deprived practically all Russian refugees of nationality and any protection of the Government of the Union of Socialist Soviet Republics. Beginning in 1922, so-called Nansen certificates were issued in order to provide some sort of identity document and various legal protection services *et al.* under international law and the League of Nations. In 1922, Fridtjof Nansen was awarded the Nobel Peace Prize for his pioneering refugee protection work. In 1930, he died and the League established the Nansen International Office for Refugees to carry on his work. For more information, see, e.g., J. E. Hassell, Russian Refugees in France and the United States between the World Wars, 1991, p. 16 *et subseq.*; M. Hansson, The Refugee Problem and the League of Nations, 1938, A.27.1931, Nansen International Office for Refugees, Report by the Governing Body to the Twelfth Assembly of League of Nations; see also M. Hansson, Former President of the Governing Body of the Nansen International Office for Refugees, Report on the Activities of the Office from July 1 to December 31, 1938, (A.19.1939. xii, 1939. xii. B.2), Statutes of Nansen International Office for Refugees as approved by the Council on January 19, 1931, League of Nations, Official Journal, February 1931, p. 309 *et subseq.*

UNHCR and other relevant stakeholders, could develop a guiding framework or instrument in this regard."[64]

The Nansen Principles seek specifically to guide responses to the challenges raised by displacement in the context of climate change and underscore the need for a human rights-based approach (Principle I). Principle VII emphasizes that "the existing norms of international law should be fully utilized" and at the same time recognizes that "normative gaps [need to be] addressed."[65]

The impetus for convening this conference came from the Bellagio expert meeting, which had concluded that there was a need "to develop a global guiding framework or instrument to apply to situations of external displacement other than those covered by the 1951 Convention, especially displacement resulting from sudden-onset disasters."[66] As it was increasingly becoming apparent that the UNHCR Ministerial Conference would fail to elaborate such a Global Guiding Framework, Switzerland and Norway in 2012 launched the Nansen Initiative, building on the principles adopted earlier on (i.e. the Nansen Principles) and the underlying political support. Whereas the UNHCR's initiative intended to cover both internal and cross-border displacement, but had not been met with consensus, the objective of the Nansen Initiative was ultimately to improve the protection of persons displaced internationally, across international state borders, by natural disasters *et al.*, through the development of a protection agenda, that is "a consensus on key principles and elements regarding the protection of persons displaced across borders in the context of natural disasters that sets the agenda for future action at domestic, regional and international levels."[67]

This Protection Agenda, which to date has been endorsed by over 100 countries,[68] is based on the following three core pillars: (i) international cooperation and solidarity between States and with the international community before (preparedness), during (admission; temporary protection), and after (durable solutions) a natural disaster; (ii) standards for the treatment of people displaced across borders regarding admission, stay, status and transition to solutions; and (iii) operational responses, including in the areas of preparedness, cross-border assistance, solutions and the respective roles of relevant disaster management, humanitarian and development actors.[69]

64 Nansen Conference, Climate Change and Displacement in the 21st Century, 2011.

65 See Nansen Initiative: A Protection Agenda for Disaster-Induced Cross-Border Displacement, Concept Note, 2012.

66 UNHCR, Summary of Deliberations on Climate Change and Displacement, 2011, p. 1.

67 Nansen Initiative, A Protection Agenda for Disaster-Induced Cross-Border Displacement: Concept Note, 2012, p. 4.

68 See www.nanseninitiative.org/more-than-100-governments-affirm-broad-support-to-better-protect-people-displaced-across-borders-by-disasters-and-the-effects-of-climate-change/ (accessed April 22, 2019).

69 Nansen Initiative, A Protection Agenda for Disaster-Induced Cross-Border Displacement, Concept Note, 2012; Nansen Initiative, Agenda for the Protection of Cross-Border Displaced Persons in the

Although not producing new legal regulations, the initiative was pioneering in that it was the first intergovernmental process to work specifically on protecting people displaced across borders by disasters and in the context of climate change. Its focus on exploring how existing political and normative frameworks could bridge the protection gap provides a solid foundation for further work.

Of importance in this respect is that the Nansen Initiative's Protection Agenda stresses the urgent need to find lasting solutions for cross-border disaster displaced persons. Effective practices in facilitating migration with dignity are referred to as a potentially positive way to cope with the effects of natural hazards, environmental degradation and climate change, including planned relocation in respect for people's rights.[70]

Finally, the Protection Agenda seeks States' commitment to enhance the protection of and assistance to IDPs in disaster contexts, by, inter alia, asking them to review domestic legislation or policies on internal displacement, in order to identify whether the notion of IDPs includes those displaced in disaster contexts. Where this is not the case, States should consider expanding their definition of IDPs in order to bring it in line with the UN Guiding Principles on Internal Displacement and relevant (sub)regional instruments. While not explicitly mentioned, this provision refers to the Kampala Convention and leaves room to consider other frameworks as well. States are also required to identify and make adequate provisions addressing disaster-related internal displacement among many other aspects.[71]

In its final part, the Protection Agenda addresses specific priority areas for future action, from collecting data and enhancing knowledge on cross-border disaster displacement, enhancing the use of humanitarian protection measures to strengthening the management of disaster displacement.[72] All of this is outlined in general terms, and more concrete implementation on how to handle cross-border disaster displacement has been formulated as the objective ahead.[73]

While the Nansen Initiative formally ended in October 2015,[74] on May 23, 2016, on the occasion of the World Humanitarian Summit in Istanbul, the so-called Platform on Disaster Displacement was launched aiming at implementing the recommendations of the Protection Agenda, which was perceived as the next key step. By virtue of this new Platform on Disaster Displacement, a new body was established, which officially started its work right after the summit in July 2016 under

Context of Disasters and Climate Change, Vol. 1, 2015, p. 15 *et subseq.*; see also *Ibid.*, Vol. 2, 2015, for regional dynamics, projections and prognostics.

70 Nansen Initiative, Agenda for the Protection of Cross-Border Displaced Persons in the Context of Disasters and Climate Change, Vol. 1, 2015, p. 37 *et subseq.*

71 *Ibid.*, p. 39 *et subseq.*

72 *Ibid.*, p. 44 *et subseq.*

73 *Ibid.*, p. 44.

74 See, e.g., www.nanseninitiative.org (accessed May 17, 2019); www.eda.admin.ch/eda/de/home/aussenpolitik/menschenrechte-menschliche-sicherheit/humanitaere-politikderschweiz/nansen-initiative.html (accessed June 11, 2019).

the Platform's steering group chairmanship of Germany and Bangladesh as vice-chair. Both expressed their commitment to protect and assist people displaced by disasters.

The original steering group of the Nansen Initiative expanded in number from nine to eighteen States,[75] representing both the Global South and the North. While the Platform has set out its priorities for the next three years (2016–2019), it is not yet clear how the Protection Agenda will be put into practice.

For many affected countries like SIDS and their vulnerable populations, it is essential to see this process gain in ambition and urgency. Fiji's Prime Minister Josaia Voreqe Bainimarama for instance said in Istanbul: "I assure the world that we in Fiji will welcome the people of Kiribati and Tuvalu as Pacific brothers and sisters with open arms. But we will need the assistance of the global community to do so. And the search for a proper framework begins now."[76]

The merit of the Nansen Initiative undeniably is to have garnered support for addressing cross-border climate displacement. It however fails to address the heart of the problem, which is internal rather than international displacement. Whereas the move away from climate change – which as a matter of fact is nowhere to be found in the Protection Agenda – allowed to rally a bigger number of States behind this initiative, the limitation to disasters raises issues whether or not the agenda would be applicable to slow-onset environmental degradation. Instead of focusing on climate change, the initiative is rather predicated on the notion of disaster as a displacement trigger. The climate displacement scope is sufficiently broad to encompass those fleeing extreme weather events as well as those displaced as a result of slow-onset environmental changes and related events. It proves to be far more difficult however to argue the inclusion of slow-onset events within the disaster scope, and it is particularly complex to determine that displacement as a result of slow events (i.e. incremental climatic change) was indeed forced and not voluntary.[77] Accordingly, the initiative's concept note outlined that slow-onset disasters are covered only "in their end phase (i.e. when droughts or rising sea levels create life-threatening situations), [as] their effects may be very similar to those of sudden-onset disasters. The relevant distinction should not be the character of the disaster but rather when it triggers displacement understood as forced movement of persons as opposed to voluntary migration."[78] This basically excludes those who moved due to

75 For the composition of the Steering Group, see http://disasterdisplacement.org/the-platform/our-architecture/ (accessed May 17, 2019); for the initial composition, see www.nanseninitiative.org/secretariat/ (May 17, 2019).

76 Prime Minister Bainimarama's Remarks at the Side Event, Commitments and Opportunities to Implement the Nansen Initiative Protection Agenda, May 25, 2016.

77 R. Zetter, Protecting Forced Migrants, A State of the Art Report of Concepts, Challenges and Ways Forward, 2014, p. 73.

78 Nansen Initiative, A Protection Agenda for Disaster-Induced Cross-Border Displacement, Concept Note, 2012; Nansen Initiative, Agenda for the Protection of Cross-Border Displaced Persons in the Context of Disasters and Climate Change, Vol. 1, 2015, p. 7.

environmental degradation processes such as droughts and desertification, increased salinization or rising sea levels, as a result of a rational decision and a form of adaptation.

## III INTERMEDIATE REMARKS

Existing frameworks such as the UN Guiding Principles on Internal Displacement, or the UN Principles on Housing and Property Restitution for Refugees and Displaced Persons, better known as the Pinheiro Principles, provide some guidance to address situations of climate displacement by means of analogy. They were however not specifically designed to handle such situations, and the Pinheiro Principles' focus is tailored to a very specific subset of the rights of displaced persons, and consequently do not comprehensively address the intricacies of climate displacement.

The Cancún Agreements sparked a number of international initiatives. The UNHCR spearheaded the attempts to elaborate a Global Guiding Framework on climate displacement, which failed owing to disagreement to tackle internal climate displacement over concerns that this could lead to a further erosion of State sovereignty. As a result, Norway and Switzerland launched the Nansen Initiative, which in turn only dealt with cross-border displacement and resulted in the adoption of the Protection Agenda. Its implementation is now being pursued within the framework of the Platform on Disaster Displacement supported by the German Government. The fact that most migration will happen internally suggests however that the bearing of these responses to address cross-border displacement will be limited.[79] In spite of existing frameworks and initiatives, there remains a lack of guidance for States to address internal climate displacement. Such guidance is also required in order to ensure a coherent as well as a coordinated and human rights-based approach for all actors working on the various aspects of climate displacement.

Climate-related human rights concerns are manifold, as its consequences may impact, among others, the rights to life, health and education as well as access to safe drinking water and to an adequate standard of living, including housing.[80] In this context, it is noteworthy that in the humanitarian and development spheres the needs-centered approach has largely been replaced by a rights-based approach.[81]

79 S. C. McAnaney, Sinking Islands? Formulating a Realistic Solution to Climate Change Displacement, 2012, p. 1174.

80 For a comprehensive analysis of the effects of climate change on human rights, see A/HRC/10/61.

81 UN Statement of Common Understanding on Human Rights-Based Approaches to Development Cooperation and Programming, http://hrbaportal.org/the-human-rights-based-approach-to-development-cooperation-towards-a-common-understanding-among-un-agencies (accessed May 4, 2019); see also, e.g., E. Cameron, Human Rights and Climate Change: Moving from an Intrinsic to an Instrumental Approach, 2010, p. 673 *et seq.*; OHCHR, Applying a Human Rights-Based Approach to Climate Change Negotiations, Policies and Measures, Guidance Note, 2010; HR/PUB/06/8, OHCHR, Frequently Asked Questions on a Human Rights-Based Approach to Development Cooperation, 2006.

This sets the frame for an ensuing consistent and anticipatory rights-based climate displacement policy and action.[82] A human rights-based approach links the assessment of the impact of climate change on individuals and groups with the accountability framework based on the human rights obligations of States to respect, protect and fulfill human rights. It is imperative that a human rights-based approach also underpins responses to climate displacement. There is a need for concrete guidance in this regard to ensure that those affected by climate displacement are placed at the center of decision-making.[83]

82 W. Sachs, Climate Change and Human Rights, 2008, p. 332 *et seq.*
83 A/RES/71/1 of September 19, 2016, para. 1; SG/A/1712 of March 9, 2017.

# 4

# A Framework for Handling Climate Displacement: The Peninsula Principles

## I INTRODUCTORY REMARKS

It might be valuable to start briefly with my own preluding research story, which commenced as of September 1, 2009, when Displacement Solutions hosted my research project on an important element of rights redress and remedy, namely the protection of the rights to housing, land and property (HLP) of refugees and all otherwise displaced persons.[1] The idea was also to work out the global legal *human rights protection texture* in place for displaced persons at large with regard to the protection of the rights to HLP, to restitution and in case of impossibility as a legal notion, to restitution in lieu. The present analysis focused on the only two existing authoritative sets of standards (i.e. the UN Guiding Principles on Internal Displacement adopted in 1998 and the UN Principles on Housing and Property Restitution for Refugees and Displaced Persons (Pinheiro Principles) endorsed in 2005, which were examined in the Chapter 3) in order to assess also how far *climate displaced persons* are granted human rights protection.

Besides, my research project inscribes into a larger bundle of analytical assessments, analysis and fact-finding efforts and activities, including on real case scenarios, led by Displacement Solutions and its director, Scott Leckie. Among these was an actionable project that emerged out of the necessity to concretely address the issue of climate affected persons on the ground together with local NGOs, grassroots organizations and international experts.

Ever since its establishment in 2006,[2] Displacement Solutions has been working on climate displacement and disposes consequently over a distinguished expertise

[1] See announcement, "New DS Project – A Legal Commentary on the Pinheiro Principles," http://displacementsolutions.org/?s=DS+project.

[2] Displacement Solutions was founded on December 28, 2006 as an NGO "to assist in resolving all forms of displacement through a reliance on human rights law. The organization has focused increasingly on newer forms of displacement, in particular the movement of people from their homes and lands due to the consequences of climate change, and over time has developed into a leading voice in this field."

on the subject matter. It became clear that the available sets of standards and human rights principles were in some way predetermined and geared toward conflict-induced displacement at least in terms of perceptional and practical application. Furthermore, it emerged that capturing the specifics as well as the general requirements of all forms of contemporary climate displacement required reconfiguration, in order to provide for a juridical sound way. Thus, the normative-legal, technical and policy guidance as well as the regulatory restatement of the current international and human rights law with respect to climate displacement would have to feed into the existing sets of standards to provide targeted guidance and also to foil ambiguous understandings that would lead to potentially deficient implementation.

The urgent need to address climate displacement through a dedicated set of principles and norms that restate international and human rights law therefore became a priority and was emphasized by grassroots organizations, international NGOs and experts.

Ultimately, Displacement Solutions, together with international donors, addressed this desideratum and initiated the Climate Displacement Law Project[3] aimed at providing comprehensive, clear and efficient guidance on climate displaced persons, deeply grounded in existing international human rights instruments and standards as well as related laws. The Director of Displacement Solutions, Scott Leckie, stated in that regard:

> The time for concerted action to prevent and resolve climate displacement is upon us, and has been for some time. In Bangladesh, the Solomon Islands, Kiribati, Panama, the US state of Alaska, Myanmar, Vietnam, Tuvalu and beyond, millions of people are facing and experiencing displacement as a result of climate change. [Displacement Solutions] has seen the human face of climate displacement up close in more than a dozen such countries, and our work to draw attention to the plight of the growing number of individuals, households or communities under threat has – out of necessity – expanded considerably since we began working on these issues in 2007.[4]

The Climate Displacement Law Project was a staggered multistage project before it reached the phase where a first draft set of principles underwent a global consultative process. One of the first objectives of the project was to develop a consolidated, globally relevant standard on climate displacement to enable governments to guide policy and legal initiatives.

See, Displacement Solutions, The Climate Displacement Land Initiative, Climate Justice for the Climate Displaced, 2016, p. 5.

3 See http://displacementsolutions.org/ds-initiatives/climate-change-and-displacement-initiative/climate-displacement-law-project/ (accessed February 2, 2019).

4 See The Peninsula Principles on Climate Displacement within States, August 18, 2013, Welcome to the Peninsula Principles on Climate Displacement within States!, p. 7, http://displacementsolutions.org/peninsula-principles/ (accessed February 2, 2019).

The declared objective was the filling of global gaps in the protection of climate displaced people and communities, and even more importantly, to ensure that concrete assurances can be provided to climate displaced persons, both within their own borders and beyond, in order to ensure that the full spectrum of their rights will be recognized, including their HLP rights.[5]

During the ensuing yearlong generalized global consultative and research process, which built the foundations for a new framework for climate displaced persons and preceded the specific, consultative, advice-receiving review process, Displacement Solutions, as a quasi-global secretariat, collected comments, questions and remarks on different iterations of the draft Peninsula Principles.[6]

It is pivotal to recall that the Peninsula Principles on Climate Displacement within States are born out of a necessity to cope with climate displacement realities, since the process was driven by people and communities claiming the protection of their rights in the wake of both large- and small-scale threats from an increasingly hostile environment.[7] Displacement Solutions merely took on the grassroots organizations' quest for guidance and solutions, notably a dedicated normative, institutional and implementation framework, and helped to facilitate and steer a process geared toward addressing the critical question of climate displacement.

The five-year-long process of broad-based consultations with thousands of people in heavily affected countries was conducted prior to a round of public soundings on the first draft.[8] The question of what kind of legal-normative instrument should be worked on, from thematic handbooks, to guidelines and draft conventions, was on the table from the beginning and was discussed ad nauseum. But there was a strong sense that if any lawmaking and norm- and standard-setting work had to be done, it should happen in the appropriate forum led by States rather than international experts, NGOs and civil society actors.

The involvement of Leckie, also an experienced international human rights lawyer who regularly advises governments and international organizations on sensitive human rights matters such as HLP restitution issues, be it in the context of peacekeeping operations or at an early stage in peace negotiations, was instrumental for the initiative, as it helped shape the approach and generated authoritative

5 See http://displacementsolutions.org/ds-initiatives/climate-change-and-displacement-initiative/climate-displacement-law-project/ (accessed February 2, 2017).

6 See S. Leckie, Land Solutions for Climate Displacement, 2014, in particular p. 1 *et subseq.* and p. 341 *et subseq.*

7 See, e.g., Displacement Solutions, Activity Reports 2007/2008, March 30, 2009; Displacement Solutions, Activity Reports 2009/2010, April 11, 2011; Displacement Solutions, Climate Change Displaced Persons and Housing, Land and Property Rights: Preliminary Strategies for Rights-Based Planning and Programming to Resolve Climate Induced Displacement, 2010.

8 The Peninsula Principles on Climate Displacement within States, August 18, 2013, Welcome to the Peninsula Principles on Climate Displacement within States!, p. 7. http://displacementsolutions.org/peninsula-principles/ (accessed March 2, 2017); H. Kramer, V. Dzihic, Die Kosovo Bilanz: Scheitert die internationale Gemeinschaft, 2005, p. 42 *et seq.*

discernment within the circle of actors on the need to engage on the issue.[9] Displacement Solutions hence acted as the vehicle for an existing inclination among donor and affected States as well as other stakeholders working on the subject to engage in a practical and simple law restatement exercise along with a documented dialogue on collecting good practices and sharing experiences in *handling climate displacement*.

The consultative process, which preceded the concrete elaboration of the set of standards and policies, attempted to engage as many individuals as possible, directly or indirectly, through their representative organizations and various institutional embodiments of their respective States. The approach, moreover, to seek feedback on all iterations of the proposed draft text from the public at large was innovative.

Progress in the initial stage depended largely on the experience and expert negotiation skills of Leckie.[10] The success in crafting the Peninsula Principles was also owing to his leadership style, drawing on his experience as facilitator of highly politicized questions of return and restitution of refugees in practically all hot spots since the end of the Cold War, negotiating HLP rights protection mechanisms in the Balkans, particularly successful in Kosovo, in East Timor, Iraq, Georgia and South Ossetia, and also in the Philippines and South Africa, as well as in Israel and Palestine and most recently in Myanmar, which often led to the recognition of rights and remedies. As an expert negotiator and social entrepreneur, his commitment to this multi-stakeholder consultation and consensus building helped to foster trust among donors as well as States and to garner support for this undertaking.[11]

9 *Cf.* S. Leckie, Returning Home: Housing and Property Restitution Rights of Refugees and Displaced Persons, 2003; Displacement Solutions, Bridging the HLP Gap, The Need to Effectively Address Housing, Land and Property Rights during Peace Negotiations and in the Context of Refugee/IDP Return: Preliminary Recommendations to the Government of Myanmar, Ethnic Actors and the International Community, 2013; Displacement Solutions, Housing, Land and Property Rights and the Quest for Peace: Guidance for Peace Mediators, 2011; S. Leckie, C. Huggins, Conflict and Housing, Land and Property Rights: A Handbook on Issues, Frameworks, and Solutions, 2011; S. Leckie, Housing, Land, and Property Rights in Post-Conflict United Nations and Other Peace Operations: A Comparative Survey and Proposal for Reform, 2009; K. Hassine, Regularizing Property Rights in Kosovo and Elsewhere, 2010; S. Leckie, Leader of the Pack: Who Will Take the Lead on Post-Conflict HLP Issues," 2009, p. 95 *et seq.*

10 Scott Leckie is also the founder of the Centre on Housing Rights and Evictions (COHRE), the first specialized international NGO working on economic, social and cultural rights.

11 *Cf.* K. Hassine, S. Leckie, The United Nations Principles on Housing and Property Restitution for Refugees and Displaced Persons: A Commentary, 2016, p. 83 *et subseq.* and p. 162 *et subseq.*; Displacement Solutions, Bridging the HLP Gap, The Need to Effectively Address Housing, Land and Property Rights during Peace Negotiations and in the Context of Refugee/IDP Return: Preliminary Recommendations to the Government of Myanmar, Ethnic Actors and the International Community, 2013, e.g. Section III. para. 15 *et subseq.*; Displacement Solutions, Housing, Land and Property Rights and the Quest for Peace: Guidance for Peace Mediators, 2011, p. 8 *et subseq.*; S. Leckie, C. Huggins, Conflict and Housing, Land and Property Rights, A Handbook on Issues, Frameworks, and Solutions, 2011, p. 92 *et subseq.*; K. Hassine, Regularizing Property Rights in Kosovo and Elsewhere, 2010, p. 23 *et seq.* and p. 53 *et seq.*; S. Leckie, Housing, Land, and Property Rights in Post-Conflict United Nations and Other Peace Operations, A Comparative Survey and

Leckie's power of persuasion was essential to convince André Hoffmann,[12] a philanthropist and outstanding personality, to join forces with him and ensure funding of the largest share of the required budget for this project through his Foundation Hoffmann, based in Morges near Geneva. In particular, Sylvie Leget, the director of Foundation Hoffmann, an experienced former humanitarian International Committee of the Red Cross diplomat, was instrumental in this endeavor as she was not only well versed with the issues, but also well acquainted with the sphere of international relations, in an interplay with the international conventional diplomacy, the United Nations at large, and humanitarian and human rights diplomacy as well as development cooperation relations. The Foundation's commitment and exceptional support throughout the entire process was the result of a genuine conviction of the value to support such a *boutique* NGO,[13] financially and through other means (e.g. by allowing it access to its *carnet d'adresse*). The comprehensive support allowed to meticulously plan and prepare the expert negotiations on the Peninsula Principles, including to equip the facilitator with the necessary political analysis, but also to allow cross-fertilization with other processes as well as sensitization and lobbying efforts, which proved indispensable for such a complex and ambitious Climate Displacement Law Project.[14]

Obviously, other donors who assisted this process come to fruition, notably the German Ministry for Development Cooperation (BMZ), which supported the land solutions to climate displacement project, also played a vital role in the development of the Peninsula Principles. In addition, the German government, for example, provided *detailed comments* to early drafts of the Principles, which helped inform and fine-tune the text.[15] Another key partner in this context was the Pictet Group Foundation,[16] which supports social, environmental and humanitarian causes, and as such partnered with Displacement Solutions beyond a merely pecuniary contribution to support its work. Due to the established support structure, in concert

Proposal for Reform, 2009, p. 329 *et subseq.*; UNHCR, Housing, Land and Property Rights in Post-Conflict Societies: Proposals for a New United Nations Institutional and Policy Framework, 2005.

12 Currently Vice-Chairman of the Board of Roche Holding Ltd., Basel, Switzerland. For further information, see www.roche.com/about/governance/board_of_directors/board_of_directors-hoffmann.htm (accessed March 22, 2019).

13 Boutique NGOs are referred to as relatively small and clearly arranged NGOs with a single thematic focus and lean administrative body.

14 For more information on the climate displacement law project, see http://displacementsolutions.org/ds-initiatives/climate-change-and-displacement-initiative/climate-displacement-law-project/ (accessed April 3, 2019).

15 S. Leckie, C. Huggins, Repairing Domestic Climate Displacement: The Peninsula Principles, 2016, p. xix.

16 See, e.g., Pictet Group Charitable Foundation: Newsletter Impact, No. 1, October 2014; or *Ibid.*, Impact, Tackling the disorder of our age, No. 3, June 2016, available at www.group.pictet/corporate/en/home/about_pictet/corporate_citizenship/charitable_foundation/impact_linkedin.html (accessed March 22, 2019); see also www.icrc.org/en/document/we-want-icrc-use-our-contribution-where-it-most-needed (accessed March 22, 2019) on their appreciation of the impartial and neutral approach.

with other crucial actors like the government of Liechtenstein, the International Federation of Red Cross and Red Crescent Societies, along with many others, the project benefited from a diversified and broad backing as well as an assistance basis to ensure the continuation of the work of Displacement Solutions on climate displacement.

## II BACKGROUND

### 1 *Process of Genesis*

In the beginning, there was only a vague vision to serve climate displaced persons in a solutions-oriented manner. Displacement Solutions started its activities in 2006, and the organization's mission statement is indicative of its initial orientation, which did not explicitly mention climate displacement: "Displacement Solutions is a not for profit international initiative that carries out projects in support of refugees and displaced persons designed to secure for them sustainable livelihood options within the context of protecting their housing, land and property rights."[17] This first paragraph indicates the rather specific expertise in the protection of HLP rights and restitution mechanisms, covering the specificities of property restitution and remedy processes. A proven expertise of the director Scott Leckie and his associates, who had pioneered with his former organization – the Centre on Housing Rights and Evictions (COHRE), was the design and implementation of such rights protection mechanisms at the request of UN agencies and peacekeeping operations as well as governments.[18]

The mission statement further explains that the organization "works together with and on behalf of people who have been displaced by conflict, forced migration or other human rights abuses, natural disasters, climate change or other circumstances beyond their control," and that it "assists displaced persons and refugees to find alternative solutions such as just compensation or resettlement when return-based restitution is physically impossible or if this is their wish."[19] Accordingly, the organization's strategic orientation focused on concrete, actionable legal and policy measures from conflict- to climate-induced displacement. The displacement realities in the field, such as the trend of increasing climate displacement, coupled with

[17] See Displacement Solutions' mission statement, http://displacementsolutions.org/about-ds/mission-statement/ (accessed March 22, 2019).

[18] See, e.g., K. Hassine, S. Leckie, The United Nations Principles on Housing and Property Restitution for Refugees and Displaced Persons: A Commentary, 2016; S. Leckie, C. Huggins, Conflict and Housing, Land, and Property Rights: A Handbook on Issues, Frameworks and Solutions, 2011, p. 10 *et subseq.*, p. 192 *et subseq.* and p. 222 *et subseq.*; M. Denmark, Land rights key to peace process, Myanmar Times, June 10, 2013; S. Leckie, Returning Home: Housing and Property Restitution Rights of Refugees and Displaced Persons, 2003, p. 3 *et seq.*; K. Hassine, Regularizing Property Rights in Kosovo and Elsewhere, 2010, p. 23 *et seq.* and p. 53 *et seq.*

[19] See Displacement Solutions' mission statement, http://displacementsolutions.org/about-ds/mission-statement/.

the challenges to deconstruct flight reasons, which may encompass, inter alia, conflict and climatic changes, along with the realization that new displacement scenarios such as the latter received little attention and raised a number of systemic normative, policy and practical protection gaps, were at the origin of Displacement Solutions' engagement on the issue.[20]

## 2 *Drafting History*

Scott Leckie and Professor David Hodgkinson, whereby the latter had initially advocated for a new convention or international instrument elaborating the rights of climate displaced persons, joined forces, bringing together different approaches, experience and expertise, which culminated in 2012 in the launching of the Climate Displacement Law Project.

This initiative integrated the earlier Climate Change and Displacement Initiative that had existed since 2007, therefore benefiting from an institutional memory, know-how and established partnerships, which catalyzed the consultative bottom-up process enormously, bringing together inter alia senior UN officials, judges, lawyers, jurists, leaders of NGOs and community-level practitioners with an agreed common objective of developing a new operative normative framework for climate displaced persons.

A key constitute feature was the participatory nature of the process. Such consultative processes are of strategic importance to assess the potential impacts of a regulatory or normative framework and to ensure that it is operational. It was designed as an inclusive process, involving particularly climate displacement affected communities and people, by virtue of modern communication means, such as the use of social networks, communication platforms *et al.*, but also in situ meetings. The objective was to "consult the world at large, visit every climate displacement hotspot across the globe, read every article, book and analysis ever written on climate displacement, and gauge existing policy, law and practice the world over to determine how best to approach the growing problem of people losing their homes and lands because of climate change."[21] Accordingly, the final drafting stage of the Peninsula Principles on Climate Displacement within States was preceded by a *1,000-day consultative process*. The drafting committee comprised an eminent group of jurists, legal scholars, practitioners and climate experts, who gathered in August 2013 on the Mornington Peninsula near Melbourne, Australia, to finalize the tenets and provide the Peninsula Principles with the final expert endorsement.

20 S. Leckie, C. Huggins, Repairing Domestic Climate Displacement: The Peninsula Principles, 2016, p. xviii.

21 *Ibid.*, p. xix.

## III NATURE AND STRUCTURE OF THE PENINSULA PRINCIPLES

### 1 *Phenotypic Placement*

The Peninsula Principles on Climate Displacement within States address the specific needs of climate displaced persons worldwide and aim at filling the policy, normative and operational protection gap in this respect. They identify rights and guarantees relevant to the protection of persons from climate displacement and to their protection and assistance *prior to* and *during* as well as *after* displacement (i.e. during return to the extent possible).

These tenets on domestic climate displacement claim to be one of the first, if not the first, formal policy document on *handling climate displacement*, founded on a broad-based consultation and deeply grounded in applicable international and human rights law. The Peninsula Principles can be considered a *recueil systématique* of relevant rules, notions, good practices and modus operandi as well as principles of international law and human rights standards, which circumscribe the rights of climate displaced persons within States. A further fundamental element is that the Principles contain protection and assistance provisions, which are not only consistent with, but build upon the Guiding Principles on Internal Displacement, which they contextualize, in order for them to be applied to climate displaced persons. The Principles therefore seek to overcome one of the weaknesses of the Guiding Principles (i.e. the lack of a specific preventative angle, which is required when dealing with climate displacement). Finally, the compilation of rights and obligations as well as policy and normative standards, combined with the human rights-based approach, is a restatement of applicable and valid international and human rights law and, thereby, remedies the existing fragmentation or lack of clarity.

### 2 *Basic Tenets*

The Peninsula Principles are based on a number of tenets that are recalled in the preambular part comprising twenty-four paragraphs. One of the assumptions, which determines the Principles' scope, is that "while climate displacement can involve both internal and cross-border displacement, most climate displacement will likely occur within State borders."[22] It can therefore be considered complementary to the Nansen Initiative, which leaves out the internal aspect of climate displacement. In this connection, given its focus on internal displacement, the preamble also affirms States' sovereignty, as a fundamental principle, mindful of the fact that precisely the concerns that climate displacement may be used by the international community or individual States to interfere in domestic affairs had stalled the earlier UNHCR-led process. The Peninsula Principles therefore provide the rationale for these inward

[22] Peninsula Principles on Climate Displacement within States, preambular para. 6.

looking standards "[a]cknowledging that States bear the primary responsibility for their citizens and others living within their territory, but recognising that, for many States, addressing the issue of and responding to climate displacement presents financial, logistical, political, resource and other difficulties."[23] The reasons furthermore for the need to address climate displacement are also outlined, including concerns that displacement may result "in the erosion of the rights of those affected, in particular vulnerable and marginalised groups, the loss of assets, housing, land, property and livelihoods, and the further loss of cultural, customary and/or spiritual identity"[24] and "if not properly planned for and managed may give rise to tensions and instability within States."[25]

The Principles furthermore build on the fundamental idea of climate justice. They reaffirm the precautionary principle,[26] and they are "[c]ognisant that the vast majority of climate displaced persons are not responsible for the processes driving climate change."[27] The principle of cooperation is central when referring to the fact that "climate change is a global problem"[28] and that "States should, on request by affected States, provide adequate and appropriate support for mitigation, adaptation, relocation and protection measures, and provide assistance to climate displaced persons."[29]

The key trait of the Peninsula Principles, which distinguishes it from other efforts, is the underlying human rights-based approach. This overriding consideration of a human rights-based approach is anchored in the preamble, which stipulates that the Principles are "[g]uided by the Charter of the United Nations." It further reaffirms the International Bill of Human Rights, consisting of the Universal Declaration of Human Rights, the International Covenant on Economic, Social and Cultural Rights, and the International Covenant on Civil and Political Rights, as well as the Vienna Declaration and Programme of Action. Moreover, essential rights in displacement situations also figure prominently in the preamble. The Principles, for instance, reaffirm the HLP rights of climate displaced persons (i.e. their right "to remain in their homes and retain connections to the land on which they live for as long as possible"),[30] as well as the right to

[23] *Ibid.*, preambular para. 11.

[24] *Ibid.*, preambular para. 1.

[25] *Ibid.*, preambular para. 10.

[26] W. Gullet, The Threshold Test of the Precautionary Principle in Australian Courts and Tribunals: Lessons for Judicial Review, 2005, p. 182 *et seq.*; R. von Schomberg, The Precautionary Principle and Its Normative Challenges, 2005, p. 19 *et seq.*; S. Boehmer-Christiansen, The Precautionary Principle in Germany: Enabling Government, 1994, p. 31 *et seq.*; N. Haigh, The Introduction of the Precautionary Principle in the UK, 1994, p. 231.

[27] Peninsula Principles on Climate Displacement within States, preambular para. 5.

[28] *Ibid.*, Foreword by S. Leckie, Director and Founder, Displacement Solutions.

[29] *Ibid.*

[30] *Ibid.*, preambular para. 7.

freedom of movement, (i.e. "the right of those who may be displaced to move safely and to relocate within their national borders over time").[31] This human rights-based approach offers a unique entry point in addressing the concerns of climate displaced persons by providing minimum standards of treatment flowing from human rights law.[32] The entitlement of persons affected by disasters to the respect for and protection of their human rights has also been included in the draft articles on the protection of persons in the event of disasters, which were recently adopted, on second reading, by the Drafting Committee of the International Law Commission and are now before the General Assembly for consideration.[33]

Finally, as is customary, the preamble refers to various other instruments that are of relevance in this context, that is, the UNFCCC and notably the Cancún Adaptation Framework, also acknowledging the IASC Operational Guidelines on the Protection of Persons in Situations of Natural Disasters, the Hyogo Framework for Action, the UN Principles on Housing and Property Restitution for Refugees and Displaced Persons and other relevant standards, as the Peninsula Principles incorporate a number of their tenets. Reference is also made to regional initiatives addressing internal displacement such as the African Union Convention for the Protection and Assistance of Internally Displaced Persons in Africa, and other initiatives such as the Nansen Initiative on disaster-induced cross-border displacement.

## 3 *Structure of the Principles*

The Peninsula Principles contain four operative parts. The first part sets out obligations of a general nature, while the remaining three parts deal with the different displacement stages. As such, the second part concerns climate displacement preparation and planning, followed by the actual displacement phase and, lastly, post-displacement matters.

31 *Ibid.*, preambular para. 8.

32 R. Zetter, The Role of Legal and Normative Frameworks for the Protection of Environmentally Displaced Persons, 2009, p. 406 *et seq.*

33 The International Law Commission decided, in accordance with Art. 23 of its statute, to recommend to the General Assembly the elaboration of a convention on the basis of the draft articles on the protection of persons in the event of disasters. The Assembly in its resolution 71/141 of December 13, 2016, took note of the draft articles and decided to include an item entitled "Protection of persons in the event of disasters" in the agenda of its 73rd session held in 2018. See A/CN.4/L.87. On November 13, 2018, a draft resolution entitled "Protection of persons in the event of disasters" A/C.6/73/L.26 of November 7, 2018, was adopted without a vote by the Committee at the same meeting. The Assembly discussed the topic and the draft articles, and took note of the views expressed in the debate, including on any future action thereon. The Assembly decided also to bring again to the attention of States the recommendation of the International Law Commission that a Convention be elaborated on the basis of the draft articles and the discussion will continue at the 75th session to be held in 2020.

## IV SCOPE OF APPLICATION

### 1 Rationale

The rationale of the Peninsula Principles on Climate Displacement within States has been aptly summarized by Mary Robinson, the first woman President of Ireland, and former UN High Commissioner for Human Rights, who currently heads her own foundation on climate justice: "States facing climate-related displacement within their borders require significant financial support and technical expertise to develop solutions that provide for the rights of those affected. The Peninsula Principles provide a normative framework, based on human rights, to address the rights of internally displaced people."[34] This is encapsulated in Principle 1, which stipulates the overall purpose of the tenets to "provide a comprehensive normative framework [...] within which the rights of climate displaced persons can be addressed" and "to set out protection and assistance principles [...] to be applied to climate displaced persons." It is important to note that the Principles' focus is climate displacement *within* a State and that it does not deal with cross-border climate displacement.

### 2 Complementing the Nomenclature

Grouping attempts in petrology may serve as an allegory for the status determination of the Peninsula Principles, whereby an aphanitic igneous rock is analyzed and classified according to the mineralogy of its phenocrysts, if the mineralogy of the groundmass does not allow for adequate assertion.[35] In the instant case, it is the analysis of the flight reasons that will allow to draw conclusions on the Principles' scope. The definition of climate displacement as provided in Principle 2(b) is "the movement of people within a State due to the effects of climate change, including sudden and slow-onset environmental events and processes, occurring either alone or in combination with other factors." Displacement is simply the change of location within the borders of the country. The ensuing Principle 2(c) is essential and defines the individual condition in a nuanced manner: "'Climate displaced persons' means individuals, households or communities who are facing or experiencing climate displacement." The use of *facing* as opposed to *face*, as well as *experiencing* rather than *experience* suggests that the Principles' scope includes anticipatory movement of people in the face of a particular threat. This interpretation is supported by the fact that earlier versions of the Principles used *face* and *experience* instead.[36] In other words, a concrete threat triggering displacement may be sufficient in certain situations for a person to fall within the scope of the Peninsula Principles' application.

[34] M. Robinson, Road to Paris, Science for Smart Policies, Q & A with Mary Robinson: What Is Climate Justice?, July 2014–April 2016.

[35] See, e.g., J. DuNann Winter, An Introduction to Igneous and Metamorphic Petrology, 2001, p. 17 *et seq.*, p. 29 *et seq.*

[36] Peninsula Principles' *Travaux Préparatoires*, August 2013 – on file with the author.

This will require assertion in order to distinguish the displacement from voluntary movement and for the victim to be able to invoke State protection action and measures. Abstract risks, by contrast, are below the threshold of the Principles and the notion of "facing and experiencing climate displacement."

This is in line with the definition of internally displaced persons adopted by the Guiding Principles, which define an IDP as

> persons or groups of persons who have been forced or obliged to flee or to leave their homes or places of habitual residence, in particular as a result of or in order to avoid the effects of armed conflict, situations of generalized violence, violations of human rights or natural or human-made disasters, and who have not crossed an internationally recognized State border.

The circumscription of flight having occurred *as a result of or in order to avoid the effects* implies that displacement may occur either after suffering the effects of coercive factors or in anticipation of such effects.

The competent authorities will need to determine, in each case, whether in light of the relevant factual circumstances a concrete displacement trigger existed. This provides the axioms for a sound and informed examination procedure for each case to ensure concrete action, once the threat of climate displacement has been recognized. Such action could include the issuance of a legally bound administrative decision, which would serve as a basis for relief action or, if applicable, for planned relocation as a preventive measure. Nevertheless, it is clear that the Principles' definition of climate displacement seeks to serve the requirements of administrative law, in order to delineate an abstract from a concrete threat of climate displacement. This is an essential human rights feature as it attempts at assisting rule of law-based decision-making, as opposed to arbitrary administrative decisions. Ultimately, these prognosis decisions need to be juridically verifiable and testable based on a criterion of an objectified ex ante rule. Therefore, the juridical review should be conducted not on the basis of the views of the factual and concrete decision-making authority but rather base itself on an average fictive authority and how it should have examined and evaluated the concrete situation from an *ex ante* and not an *ex post* perspective. The existence of a climate displacement threat with sufficient probability subscribes to the principle of proportionality,[37] meaning the greater the expected damage, the lower the requirements on probability and vice versa.

## 3 Restatement of International Norms and Law

Climate displaced persons have cross-cutting protection needs. Addressing them requires a collaborative response. In light of the internal nature of displacement that falls within the scope of the Peninsula Principles, national State authorities are the primary duty-bearer to ensure the protection of the rights of climate displaced persons across all phases of climate displacement.

[37] BVerfGE 3, 383, 399 (1954).

Although it could be argued that the Peninsula Principles lack binding force, as States have not signed and ratified them, their content to a large extent derives from existing international human rights and law, by contrast to the desideratum of a protection regime in the form of an international convention or any other binding instrument, providing guidance to all actors dealing with climate displaced persons, including national and international actors. The Principles serve as a reference tool consisting of norms and standards of international human rights law applicable to situations of climate displacement. They restate and reaffirm international legal standards in a coherent manner and contextualize them (i.e. adapt them to the specific protection needs of climate displaced persons).

Given that the Peninsula Principles explicitly build on the UN Guiding Principles on Internal Displacement, it is important to be mindful of the wide endorsement and approval of the Guiding Principles, in particular by States. Most significantly, the international community endorsed the Guiding Principles at the 2005 World Summit, further to their presentation to the Commission on Human Rights.[38] Besides numerous references to them at the international and national levels, the Principles earned recognition through the adoption of the African Union Convention for the Protection and Assistance of Internally Displaced Persons in Africa, the so-called Kampala Convention.[39] This was a milestone in invigorating these international norms and standards. Consequently, the UN Special Rapporteur on the human rights of internally displaced persons in 2011 concluded that "there are some indications that the Guiding Principles are emerging as customary law, providing a binding interpretation of the international legal norms upon which they are based."[40]

The Peninsula Principles therefore complement the Guiding Principles, which do not contain specific reference to climate change or climate displacement, with a comprehensive normative framework, based on principles of international law, human rights obligations and notably good practice, within which the specific rights of climate displaced persons can be managed and handled.

By applying the Peninsula Principles to internal climate displacement and calling upon States, inter alia, to prevent and avoid conditions that might lead to climate displacement, providing assistance to affected individuals, implementing national

[38] In 2005, more than 190 States adopted the World Summit Outcome document, stating, "We recognize the Guiding Principles on Internal Displacement 39 as an important international framework for the protection of internally displaced persons and resolve to take effective measures to increase the protection of internally displaced persons"; see A/RES/60/1 of September 16, 2005, para. 123; see also E/CN.4/RES/2004/55 of April 20, 2004, operative para. 6: The Commission on Human Rights "[e]xpresses its appreciation of the Guiding Principles on Internal Displacement as an important tool for dealing with situations of internal displacement, welcomes the fact that an increasing number of States, United Nations agencies and regional and non governmental organizations are applying them as a standard, and encourages all relevant actors to make use of the Guiding Principles when dealing with situations of internal displacement."

[39] See, e.g., M. Klein Solomon, K. Warner, Protection of Persons Displaced as a Result of Climate Change: Existing Tools and Emerging Frameworks, 2013, p. 267 *et subseq.*

[40] A/HRC/12/21.

climate displacement prevention programs and coordinating adaptation assistance with other States and international agencies, they aim at providing a coherent rights-based strategy throughout the entire displacement cycle.[41]

Following also the recognition and importance to use indicators for the realization of human rights, the Principles incarnate parameters to reinforce a culture of accountability on the subject, by also offering institutional planning guidance, including for climate displacement risk management, participation and consent of affected individuals, land identification, as well as post-displacement return.[42]

At the same time the value of the Peninsula Principles as a simple restatement of international law could be challenged. It could be argued indeed that in order to provide an added value, codification efforts would need to contain a progressive element. "Codification which constitutes a record of the past rather than a creative use of the existing materials – legal and others – for the purpose of regulating the life of the community is a brake upon progress."[43] There is however another justification of codification, which is the argument of necessity, notably in case of "subjects which have been regulated by international law, but the regulation of which is unsatisfactory and fragmentary."[44]

## V GENERAL OBLIGATIONS UNDER THE PENINSULA PRINCIPLES

### 1 *Prevention and Avoidance*

Pursuant to the prevention onus upon States, Principle 5 stipulates under the title *prevention and avoidance* that "States should, in all circumstances, comply in full with their obligations under international law so as to prevent and avoid conditions that might lead to climate displacement." The premise is that the State and

[41] B. Doherty, General Obligations, 2016, p. 70 *et subseq.*

[42] In 2004, the Constitutional Court of Colombia instructed the Government to provide detailed information on the rights of internally displaced persons and the policies affecting them. The Court ascertained that the Government had failed to provide sufficient resources and to create the required institutional capacity to uphold their rights. It requested quantifiable and comparable indicators to be identified and used to ensure a culture of accountability. In particular, indicators would help to assess the effectiveness of government policies, specifically the impact of its remedial measures; measure the enjoyment of rights in every phase of displacement, specifically by children, women, persons with disabilities and indigenous peoples; and focus on and provide only the most essential information, in view of the resource implications. As a result, the Government of Colombia put in place an evaluation mechanism with indicators to measure progress in the implementation of the rights of IDPs, including their rights to food, health, education, and liberty and security. See HR/PUB/12/5, Human Rights Indicators: A Guide to Measurement and Implementation, 2012, p. 3 *et subseq.* and p. 62; J. Rothring, M. Romero, Measuring the Enjoyment of Rights in Colombia, 2008, p. 64 *et subseq.*; M. J. Cepeda-Espinosa, How Far May Colombia's Constitutional Court Go to Protect IDP Rights?, 2006, p. 21 *et seq.*; Brookings Institution, Internal Displacement and the Construction of Peace, Summary Report, Bogota, Colombia, November 11–12, 2008.

[43] A/CN.4/Rev.1, 1949, para. 13.

[44] *Ibid.*

governmental agencies are assigned obligations of due diligence and a duty of care. These obligations can be infringed through negligence. The preventative focus of this tenet further implies a proactive imperative (i.e. to adopt all necessary measures to avert climate displacement). Consequently, *avoidance* in this sense implies refraining from any State-driven or responsible positive action that *might* cause climate displacement.[45]

This Principle corresponds to the *Vorsorgeprinzip* introduced into German environmental law in the early 1970s as a benchmark for government decision-making.[46] The concrete manifestations of this tenet are preventive and anticipatory measures with regard to activities that might harm the environment, conscious of a degree of uncertainty about cause–effect relations. A fortiori, precisely owing to causalities that are not fully established, this precautionary principle approach is an embodiment of the *"Do No Harm"* rule, since by stipulating *prevent* and *avoid*, there is an obligation to anticipate, shield and care to the extent possible without awaiting scientific proof. The precautionary principle notion initially encompassed preventive anticipation, safeguarding and protection of the environment as well as proportionality of the response or cost-effectiveness of margins of error to show that the selected degree of restraint was excessive. In spite of the discussion on the *climate displacement causation challenge*,[47] Principle 5 does not consider causation in any way. However, in accordance with the UNFCCC and as laid down in Principle 15 of the 1992 Rio Declaration on Environment and Development,[48] the precautionary principle stipulates: "Where there are threats of serious or irreversible damage, lack of full scientific certainty should not be used as a reason for postponing cost-effective measures to prevent environmental degradation."[49] The precautionary principle in the context of environmental protection is essentially about the management of scientific risk and a fundamental component of the concept of ecologically sustainable and sound development. Thus, in compliance with this precautionary tenet, Principle 5 is urging States to take all possible measures – from policy to regulatory actions. In simple terms, the maxim for action could also be "prevention is better than cure."

By the same token, Article 3(3) of the UNFCCC stipulates

45 See HR/PUB/12/5, Human Rights Indicators: A Guide to Measurement and Implementation, 2012, p. 3 *et subseq.* and p. 62; J. Rothring, M. Romero, Measuring the Enjoyment of Rights in Colombia, 2008, p. 64 *et subseq.*; M. J. Cepeda-Espinosa, How Far May Colombia's Constitutional Court Go to Protect IDP Rights?, 2006, p. 21 *et seq.*

46 S. Boehmer-Christiansen, The Precautionary Principle in Germany: Enabling Government, 1994, p. 31 *et seq.*; N. Haigh, The Introduction of the Precautionary Principle in the UK, 1994, p. 231.

47 S. Leckie, Using Human Rights to Resolve the Climate Displacement Problem: The Promise of the Peninsula Principles, 2016, p. 5 *et subseq.*

48 *Cf.* O. Höll, Die Ergebnisse der UN-Umweltkonferenz von Rio (3.–14. Juni 1992), 1992; O. Höll, Ökologisierung der Internationalen Politik als Konfliktpotential, Die Konferenz über Umwelt und Entwicklung von Rio und die Folgen, 1993.

49 A/CONF.151/26 (Vol. 1), Rio Declaration on Environment and Development, adopted on June 14, 1992, Principle 15.

> The Parties should take precautionary measures to anticipate, prevent or minimize the causes of climate change and mitigate its adverse effects. Where there are threats of serious or irreversible damage, lack of full scientific certainty should not be used as a reason for postponing such measures, taking into account that policies and measures to deal with climate change should be cost-effective so as to ensure global benefits at the lowest possible cost. To achieve this, such policies and measures should take into account different socio-economic contexts, be comprehensive, cover all relevant sources, sinks and reservoirs of greenhouse gases and adaptation, and comprise all economic sectors. Efforts to address climate change may be carried out cooperatively by interested Parties.

Government authorities or local authorities therefore have an obligation to anticipate, prevent and counter the causes of environmental degradation that might lead to climate displacement. Where there are threats of serious and irreversible damage, lack of scientific certainty should not be used as a reason for postponing measures to prevent environmental degradation. Furthermore, the *onus of proof* is on the actor or the developer to show that his action is environmentally friendly. Consequently, applying the precautionary principle in analogy to climate displacement requires explicit consideration of the kind and extent of potential harm, along with the degree of uncertainty about the likelihood of harm, before deciding on the measures and means of address. The dogmatic regulatory content of the precautionary principle requires consideration of potential harm and probability. Henceforth, decisions require a case-by-case approach, establishment of facts and appreciation of values prior to a decision on the merits.[50]

Full respect of relevant norms of international law, including human rights and humanitarian law, by domestic and international actors, both governmental and non-governmental, very significantly reduces the risk of internal displacement in situations of tensions and disturbances or armed conflict. Many situations of displacement could be avoided or minimized if guarantees of international human rights and humanitarian law were adequately adhered to.[51] In situations where national authorities (i.e. the official authorities of the State) lack the capacity to fulfill this duty or are no longer functional due to the particular circumstances that prevail in the country, Principle 5 of the UN Guiding Principles on Internal Displacement, which addresses the duties of all authorities, including de facto organs and international actors, becomes especially important. This is also relevant in the context of Peninsula Principle 5, as the protection of human rights means to assume the primary responsibility for protecting and assisting displaced persons and respecting, protecting and fulfilling their civil and political as well as their economic, social and cultural rights, which as a result contributes to alleviating displacement triggers. Accordingly, States have to take the legal and administrative

[50] W. Gullet, The Threshold Test of the Precautionary Principle in Australian Courts and Tribunals: Lessons for Judicial Review, 2005, p. 182 *et seq.*; R. von Schomberg, The Precautionary Principle and Its Normative Challenges, 2005, p. 19 *et seq.*

[51] W. Kälin, Guiding Principles on Internal Displacement: Annotations, 2008, p. 41.

measures necessary to effectively address situations of internal displacement. In this sense, the UN General Assembly "[e]ncourages States to continue to develop and implement domestic legislation and policies dealing with all stages of displacement, including through the identification of a national focal point within the Government for issues of internal displacement, and through the allocation of budget resources."[52] States do not have an obligation to identify and name a national focal point on internally displaced persons or climate displacement, but they have to comply with the prohibition of arbitrariness and not to contribute in any way to the displacement of persons in accordance with Principle 6(1) of the Guiding Principles on Internal Displacement. Principle 6(2)(d) further prohibits arbitrary displacement including displacement "[i]n cases of disasters, unless the safety and health of those affected requires their evacuation," which extends also in its applicability to climate displacement. This is deeply grounded in international and human rights law and follows from the right to freedom of movement and the right to choose one's residence within a State stipulated in the International Bill of Human Rights, precisely in key Article 12 of the International Covenant on Civil and Political Rights. It does not only guarantee the right to liberty of movement and the freedom to choose one's residence that includes the right to remain there, but stipulates that this right "shall not be subject to any restrictions except those which are provided by law, are necessary to protect national security, public order (*ordre public*), public health or morals or the rights and freedoms of others, and are consistent with the other rights recognized in the present Covenant" (paragraph 3).[53] The guarantees regarding the freedom from arbitrary interference with one's home (Article 17 ICCPR) contain similar limitation clauses.[54]

Accordingly, forced evictions[55] in line with international human rights law require observance of the principles of proportionality, transparency and respecting the rights framework and security of tenure as well as active, free and meaningful

[52] A/RES/62/153, paragraph 11. For a more detailed elaboration of steps to be taken by governments, see E/CN.4/2006/71/Add.1, Framework for National Responsibility, 2005.

[53] M. Nowak, UN Covenant on Civil and Political Rights: CCPR Commentary, 2005, Art. 12, para. 41.

[54] W. Kälin, Guiding Principles on Internal Displacement: Annotations, 2008, p. 26 *et seq.*; see also ICCPR, adopted on December 16, 1966, General Assembly resolution 2200A (XXI), entered into force March 23, 1976. In its general comment on Art. 12, the Human Rights Committee, the treaty body for the ICCPR, explains that "[e]veryone lawfully within the territory of a State enjoys, within that territory, the right to move freely and to choose his or her places of residence" and that forced internal displacement is prohibited, see CCPR/C/21/Rev.1/Add.9 (1999), paras. 4 and 7; M. Nowak, UN Covenant on Civil and Political Rights: CCPR Commentary, 2005, Art. 12, para. 44 *et subseq.*

[55] The obligation of States to refrain from, and protect against, forced evictions from home(s) and land arises from several international legal instruments including the UDHR, the ICESCR (Art. 11, para. 1), the CRC (Art. 27, para. 3), the nondiscrimination provisions found in Art. 14, paragraph 2(h), of CEDAW, and Art. 5(e) of the ICERD. In its resolution 1993/77, the Commission on Human Rights stated that the "practice of forced eviction constitutes a gross violation of human rights, in particular the right to adequate housing." In 1977, the CESCR issued its General Comment no. 7 on forced evictions. Evictions must be carried out lawfully, only in exceptional circumstances, and in full accordance with relevant provisions of international human rights and humanitarian law. In 2007, the Special Rapporteur on adequate housing presented to the Human Rights Council a set of basic principles and guidelines on development-based evictions and displacement. These guidelines aim to

participation. Using potential threats as a pretext to undertake climate change adaptation measures that would result in forced evictions by State authorities is thereby precluded. Practice showed already that post-disaster reconstruction or the improper use of disaster risk reduction laws or housing building standards become an excuse for evicting and displacing people from their homes.[56]

## 2 National Implementation Measures

The Peninsula Principles, alike other policy/normative frameworks, should be seen as a catalyst for concrete action, the lack of which too often constitutes the main obstacle for the actual realization of rights. The Principles provide guidance for policy formulation and evaluation to handle climate displacement within States and toward strengthening national capacity for human rights implementation, thereby addressing an existing gap.[57]

Principle 7 requires States to "incorporate climate displacement prevention, assistance and protection provisions [...] into domestic law and policies, prioritising the prevention of displacement." This Principle is key to implementation (i.e. to transform norms into action), as it sets the scene and facilitates the enactment of the framework along with its scope and structure as well as ensuing policy and legal measures. It is congruent with States' primary responsibility to handle climate displacement. It is also in line with Principle 3(1) of the Guiding Principles on Internal Displacement, which states that "[n]ational authorities have the primary duty and responsibility to provide protection and humanitarian assistance to internally displaced persons within their jurisdiction." It is nothing else than the embodiment of the principle of State sovereignty as comprised, inter alia, in Article 2(7) of the UN Charter, prohibiting the interference in matters that are essentially within the domestic jurisdiction of any State.[58] Providing protection and humanitarian assistance to nationals, including IDPs, is a primary duty and responsibility of the State. Accordingly, the UN General Assembly restated a number of times that "the sovereignty of affected States and their primary role in the initiation, organization, coordination and implementation of humanitarian assistance within their respective territories."[59]

assist States in developing policies and legislations to prevent forced evictions at the domestic level. The Basic Principles and Guidelines on Development-Based Evictions and Displacement represent a further development of the United Nations Comprehensive Human Rights Guidelines on Development-Based Displacement (E/CN.4/Sub.2/1997/7, annex). The Basic Principles and Guidelines on Development-Based Evictions and Displacement are contained in annex I of the report of the Special Rapporteur, A/HRC/4/18.

56 See UN-Habitat/UNOHCHR, Forced Evictions, Fact Sheet No. 25, Rev.1, 2014, p. 2 *et seq.*; see also COHRE, Sources No. 4: Legal Resources for Housing Rights: International and National Standards, 2000.

57 See HR/PUB/12/5, Human Rights Indicators: A Guide to Measurement and Implementation, 2012.

58 G. Nolte, Art. 2(7), para. 10 *et seq.*, in: B. Simma *et al.*, The Charter of the United Nations, Vol. 1, 2012.

59 A/RES/45/100 of December 14, 1990; see also, e.g., A/RES/46/182 of December 19, 1991. At the regional level, the content of Principle 3 has been explicitly recognized in the Great Lakes IDP Protocol (Art. 3(3)) and by the Council of Europe, Committee of Ministers, Rec(2006)6, paragraph 4; see also, e.g., UNFCCC in its preambular para. 9 in recognition of the *principle of sovereignty of States*.

By way of illustration, the Peninsula Principles have been instrumental in defining the scope and direction of Bangladesh's National Strategy for Disaster and Climate Induced Internal Displacement dated September 13, 2015. This constituted a litmus test for the acceptance, rights compliance, functionality and function ability of the Principles as they formed the basis for the adoption of Bangladeshi directive policies. The National Strategy on the Management of Disaster and Climate Induced Internal Displacement (NSMDCIID) under the Comprehensive Disaster Management Programme (CDMP II) of the Ministry of Disaster Management and Relief (MoDMR) based on recent estimates cited in this Strategy suggests that, by 2050, one out of seven Bangladeshis will be displaced by climate change. The estimates of Bangladesh's Internal Displacement Monitoring Centres show that 4.7 million people were displaced due to disasters between 2008 and 2014. By 2080, it is likely that 13 percent of Bangladesh's coast will have disappeared owing to rising sea levels.[60] In light of this, Bangladesh's NSMDCIID, which was developed with the support and endorsement of the European Union, Norway, Sweden and the UNDP, as well as UK Aid and Australian Aid, bases itself on the Peninsula Principles, and exhibits already the existence of a certain degree of international acceptance.[61] It is of interest that the rationale of the strategy includes the adoption of a human rights-based approach, since the strategy lists the rights affected and the need to comply with and to protect human rights.[62] The basis for action grounds in the following discernment: "Since strategic responses cannot simply be assumed to be a neutral activity affecting everyone equally and in a positive way, any viable approach to displacement should be based on the principles of human rights enshrined in the international and regional human rights instruments."[63]

## 3 *International Cooperation and Assistance*

Principle 8 calls on States to "cooperate in the provision of adaptation assistance (to the maximum of their available resources) and protection for climate displaced persons"

60 J. Pender, Community-Led Adaptation in Bangladesh, 2008, p. 54 *et subseq.*

61 See Ministry of Disaster Management and Relief (MoDMR), National Strategy on the Management of Disaster and Climate Induced Internal Displacement (NSMDCIID), 2015; Displacement Solutions, Peninsula Principles highlighted in Bangladesh's National Strategy for Disaster and Climate Induced Internal Displacement, December 2, 2015.

62 Ministry of Disaster Management and Relief (MoDMR), National Strategy on the Management of Disaster and Climate Induced Internal Displacement (NSMDCIID), 2015, p. 5, "Affected persons may face multiple human rights challenges ... they experience gender-based violence, unequal access to assistance, basic goods and services and discrimination in aid provision; abuse, neglect and exploitation of children; family separation, particularly for children, older persons, persons with disabilities and other individuals who may rely on family support for their survival; loss/destruction of personal documentation and difficulties to replace it, in particular due to inadequate birth registration mechanisms; inadequate law enforcement mechanisms and restricted access to a fair and efficient justice system; lack of effective feedback and complaint mechanisms; unequal access to employment and livelihood opportunities; forced relocation; unsafe or involuntary return or resettlement of persons displaced by the disaster; or lack of property restitution and access to land."

63 *Ibid.*, p. 9.

since climate displacement "is a matter of global responsibility." This is grounded in the conception of an international community of States compelled to cooperate and collaborate with each other on global issues and, in the present case, in the handling of climate displacement. The International Bill of Human Rights and the core human rights treaties, notably the 1966 ICESCR, the Convention on the Rights of the Child, and the Convention on the Rights of Persons with Disabilities, ascertain international cooperation and assistance as essential and required to facilitate the progressive realization of economic, social and cultural rights.[64]

The principle of international cooperation is moreover enshrined in the UN Charter. Article 1(3) provides one of the purposes of the organization is "[t]o achieve international co-operation in solving international problems of an economic, social, cultural, or humanitarian character, and in promoting and encouraging respect for human rights and for fundamental freedoms for all without distinction as to race, sex, language, or religion."[65] Article 56 of the UN Charter furthermore asserts that "[a]ll Members pledge themselves to take joint and separate action in co-operation with the Organization for the achievement of the purposes set forth in Article 55."[66]

Similarly, the ICESCR, in Article 2(1), affirms that "[e]ach State Party to the present Covenant undertakes to take steps, individually and through international assistance and co-operation, especially economic and technical, to the maximum of its available resources, with a view to achieving progressively the full realization of the rights recognized in the present Covenant by all appropriate means, including particularly the adoption of legislative measures."

Also relevant here is that Article 22 of the UN Charter, which clarifies the following:

> The Economic and Social Council may bring to the attention of other organs of the United Nations, their subsidiary organs and specialized agencies concerned with furnishing technical assistance any matters arising out of the reports referred to in this part of the present Covenant which may assist such bodies in deciding, each within its field of competence, on the advisability of international measures likely to contribute to the effective progressive implementation of the present Covenant.[67]

Besides, Article 23 of the ICESCR states: "The States Parties to the present Covenant agree that international action for the achievement of the rights recognized in the present Covenant includes such methods as the conclusion of conventions, the adoption of recommendations, the furnishing of technical assistance and the holding of regional meetings and technical meetings for the purpose of consultation and study organized in conjunction with the Governments concerned."

64 See, e.g., ICESCR, Art. 2(1); CRC, Preamble, para. 11 and Art. 4; CRPD, Art. 4(2).
65 R. Wolfrum, Art. 1, para. 7 *et seq.*, in: B. Simma *et al.*, The Charter of the United Nations, Vol. 1, 2012.
66 T. Stoll, Art. 56, para. 4 *et seq.*, in: B. Simma *et al.*, The Charter of the United Nations, Vol. 2, 2012.
67 D.-E. Khan, Art. 22, para. 1 *et seq.*, in: B. Simma *et al.*, The Charter of the United Nations, Vol. 1, 2012.

The Peninsula Principles subscribe themselves fully to the aforementioned and reiterate the principle of cooperation among the tenets' general obligations. Cooperation has several facets. It is not only an obligation of the international community (i.e. between States providing assistance), cooperation is also stipulated as a right of the State concerned. Principle 8 provides that "States have the right to seek cooperation and assistance from other States and relevant international agencies" in fulfilling their obligations to prevent and respond to climate displacement. This right is coupled with a duty to accept appropriate assistance and support from other States and relevant international agencies, whether made individually or collectively, if the State concerned is unable to adequately prevent and respond to climate displacement. The Principle moreover urges the international community – of utmost importance thereby is South–South cooperation[68] – to provide cooperation and, if needed, assistance to ensure that climate displacement-affected States can cope with their responsibilities and obligations. International environment law, such as the UNFCCC, but also the 1951 Refugee Convention, formulate similar obligations for collaboration and assistance, namely if the capacity of the concerned State is limited in coping with its obligations.[69] Notably, this cooperation can also be extended and directed toward contributing to ensuring the rights protection of communities and people affected by climate displacement.

This also ties up with the 1993 Vienna Declaration and Programme of Action adoption, which marked a turning point in human rights promotion and protection, including for the issue of displacement as topical priority. Paragraph 23 of the Vienna Declaration and Programme of Action "recognizes that gross violations of human rights [. . .] are among the multiple and complex factors leading to displacement of people."[70] This has to be read in conjunction with the preamble, which reaffirms "the commitment contained in Article 56 of the Charter of the United Nations to take joint and separate action, placing proper emphasis on developing effective international cooperation for the realization of the purposes set out in Article 55, including universal respect for, and observance of, human rights and fundamental freedoms for all." Furthermore, the Vienna Declaration and Programme of Action continues by expressing its resolve as it states: "Determined to take new steps forward in the commitment of the international community with a view to achieving substantial progress in human rights endeavours by an increased and sustained effort of international cooperation and solidarity."[71] The enhancement of international cooperation in the field of

68 *Cf.* 2030 Agenda for Sustainable Development and its 17 Sustainable Development Goals (SDGs), including the unequivocal call to *leave no one behind* (A/RES/70/1 of September 25, 2015); see also Sustainable Development Knowledge Platform, https://sustainabledevelopment.un.org/ (accessed May 20, 2019); United Nations Office for South–South Cooperation (UNOSSC), www.unsouthsouth.org (accessed May 20, 2019).

69 *Cf.* UNFCCC, Art. 4; 1951 Refugee Convention, Preamble.

70 DPI/1294 Rev.2, OHCHR/UNDPI, Vienna Declaration and Programme of Action, 20 Years Working for Your Rights, 1993 World Conference on Human Rights, June 2013, p. 24.

71 *Ibid.*, p. 1 *et subseq.* and p. 22 *et subseq.*

human rights is being considered essential for the achievement of the purposes of the United Nations (i.e. simply put, the full realization of peace and security, development and human rights for all, everywhere).[72]

The practical purpose of the cooperation provision is to ensure the availability of means, expertise and resources to enable and foster the Principles' implementation, especially because the States most affected by climate displacement may be lacking the means to address the related challenges.

## VI HUMAN RIGHTS IMPERATIVES

### 1 Systematic Footing

The Peninsula Principles' underlying assumption is that adopting a human rights-based approach will enable the State in its various embodiments and by virtue of its various means of action – from normative, policy to even polity measures – to place the handling of domestic climate displacement on a more systematic footing that is anchored in rights and not charity-driven, and thus not merely reactive but proactive, enabling preventive measures and participatory approaches.

According to Mary Robinson, former United Nations High Commissioner for Human Rights, "A rights-based approach is a conceptual framework for the process of human development that is normatively based on international human rights standards and operationally directed to protecting and promoting human rights. The rights-based approach integrates the norms, standards and principles of international human rights system into plans, policies and processes of development."[73] Following from this, systematic human rights footing of all decisions, activities and processes as well as policy and development as called for by the UN Declaration on the Right to Development and agreed upon in the UN Common Understanding of a Human Rights-Based Approach to Development Cooperation, means the integration and ideally internalization of the human rights-based approach.[74] The UN Common Understanding emphasizes key human rights

[72] R. Wolfrum, Art. 1, para. 7 *et seq.*, in: B. Simma *et al.*, The Charter of the United Nations, Vol. 1, 2012; T. Stoll, Art. 56, para. 4 *et seq.*, *Ibid.*, Vol. 2, 2012.

[73] M. Robinson, What Rights Can Add to Good Development Practice, 2005, p. 38.

[74] In 1997, the UN Secretary-General called on all entities of the UN system to mainstream human rights into their various activities and programs within the framework of their respective mandates. Since then a number of agencies adopted a human rights-based approach to their development cooperation, but each had its own interpretation. A common understanding was reached in 2003 and adopted by the United Nations Development Group (UNDG). See http://hrbaportal.org/the-human-rights-based-approach-to-development-cooperation-towards-a-common-understanding-among-un-agencies (accessed May 4, 2019); see also, e.g., E. Cameron, Human Rights and Climate Change: Moving from an Intrinsic to an Instrumental Approach, 2010, p. 673 *et seq.*; OHCHR, Applying a Human Rights-Based Approach to Climate Change Negotiations, Policies and Measures, Guidance Note, 2010; HR/PUB/06/8, OHCHR, Frequently Asked Questions on a Human Rights-Based Approach to Development Cooperation, 2006.

principles like, in particular, the following conceptual fundamental tenets: universality and inalienability, indivisibility, interdependence and interrelatedness, non-discrimination and equality, participation and inclusion, accountability and the rule of law.[75]

It sketches a conceptual and structured framework for development, based on international human rights standards, in order to promote and protect human rights in the context of all development activities. A rights-based approach examines duties and responsibilities, detects disparities, identifies circumstances and situations of fragility and even frailty, and pursues to repair and redress, for instance, against direct and indirect discriminatory practices and inequality settings resulting in exclusion. Systematic human rights footing thus provides the common ground for anchoring plans, policies and programs in a system of rights that is based on the rule of law. In turn, ensuing onuses as established by international and human rights law and restated in numerous standards, compilations of rights and sets of principles are conveyed in normative clarity and certainty.

The critical characteristics of a human rights-based approach are that policies and programs are formulated in line with human rights, identifying human beings as rights-holders along with their entitlements. Accordingly, the corresponding duty-bearers and their obligations need to be clearly determined, so as to allow for ways, methods and possibilities to strengthen the capacities of rights-holders to make their claims, and those of duty-bearers to meet their obligations.

**Applying a Human Rights-Based Approach to Climate Change Negotiations, Policies and Measures**

The Human Rights Council noted that "climate change-related impacts have a range of implications, both direct and indirect, for the effective enjoyment of human rights" and recalled "that in no case may a people be deprived of its own means of subsistence." It affirmed that "human rights obligations and commitments have the potential to inform and strengthen international and national policymaking in the area of climate change," thereby noting the importance of applying a human rights-based approach to the climate change.

Source: Human Rights Council resolution 10/4 of March 25, 2009.

In light of these characteristics, principles and standards derived from international human rights law – especially from the Universal Declaration of Human Rights (UDHR) and the core human rights treaties – should guide all policies and programming in all phases of the process.

75 See http://hrbaportal.org/the-human-rights-based-approach-to-development-cooperation-towards-a-common-understanding-among-un-agencies (accessed May 4, 2019).

For that reason, systematic footing of human rights should also guide any climate change adaptation or mitigation measures, such as the promotion of alternative energy sources, forest conservation or tree-planting projects, planned relocation or resettlement schemes and others. Affected individuals and communities must participate, without discrimination whatsoever, in the design and implementation of these projects. They must have access to due process and to remedy, if their rights are being violated.[76]

In a sort of legal syllogism, trying to capture the essence and to clarify the components of a human rights-based approach, Robinson formulated the following guiding questions as subsumption support:

> In each situation we confront, a rights-based approach requires us to ask: What is the content of the right? Who are the human rights claim-holders? Who are the corresponding duty-bearers? Are claim-holders and duty-bearers able to claim their rights and fulfil their responsibilities? If not, how can we help them to do so? This is the heart of a human rights based approach.[77]

**Essential attributes of a human rights-based approach**

A rights-based approach analyses obligations, inequalities and vulnerabilities, and seeks to redress discriminatory practices and unjust distributions of power. It anchors plans, policies and programs in a system of rights, and corresponding obligations established by international law.

- As policies and programs are formulated, the main objective should be to fulfil human rights.
- The rights-holders and their entitlements must be identified as well as the corresponding duty-bearers and their obligations in order to find ways for strengthening the capacities of rights-holders to make their claims and of duty-bearers to meet their obligations.
- Principles and standards derived from international human rights law – especially the Universal Declaration of Human Rights and the core universal human rights treaties – should guide all policies and programming in all phases of the process.

Source: Submission of the Office of the High Commissioner for Human Rights to the 21st Conference of the Parties to the United Nations Framework Convention on Climate Change, p. 9.

76 For further information on the human rights-based approach, see http://hrbaportal.org/the-human-rights-based-approach-to-development-cooperation-towards-a-%20common-understanding-among-un-agencies (accessed March 20, 2019).

77 M. Robinson, former UN High Commissioner for Human Rights, Comments at 2nd Interagency Workshop on Implementing a Rights-Based Approach in the Context of UN Reform, 2003.

## 2 Nondiscrimination and Equality

The prohibition of discrimination is an overarching tenet of the Peninsula Principles, which underpins all Principles, including those listed under General Obligations, which are applicable to the various displacement phases. Principle 3 provides that

> States shall not discriminate against climate displaced persons on the basis of their potential or actual displacement, and should take steps to repeal unjust or arbitrary laws and laws that otherwise discriminate against, or have a discriminatory effect on, climate displaced persons. Climate displaced persons shall enjoy, in full equality, the same rights and freedoms under international and domestic law as do other persons in their country, in particular housing, land and property rights.

The right, not to be discriminated against – directly or indirectly[78] – forms a supreme tenet of normative value. Thus, bearing in mind a main feature of the human rights framework, namely that human rights are indivisible, interdependent and interrelated.[79]

The principle of nondiscrimination is enshrined as one of the purposes of the United Nations in Article 1(3) of the UN Charter,[80] and it is stipulated in most human rights instruments, particularly in the UDHR[81], the ICCPR,[82] as well as the ICESCR.[83] Because of its basic and general character, the principles of nondiscrimination and equality before the law and equal protection of the law are sometimes expressly restated and referred to in articles relating to particular categories of human rights.[84] The same idea is being applied here with the Peninsula Principles standard set of rules by means of analogy.

78 See, e.g., CCPR/C/74/D/919/2000, March 26, 2002; CCPR/C/67/D/666/1995, November 9, 1999.

79 See A/Conf. 157/23 of July 12, 1993, para. 5 and 18.

80 See Art. 1, para. 3 of the UN Charter: "To achieve international cooperation in solving international problems of an economic, social, cultural, or humanitarian character, and in promoting and encouraging respect for human rights and for fundamental freedoms for all without distinction as to race, sex, language, or religion"; see R. Wolfrum, Art. 1, para. 7 *et seq.*, in: B. Simma *et al.*, The Charter of the United Nations, Vol. 1, 2012.

81 See Art. 2 of the UDHR: "Everyone is entitled to all the rights and freedoms set forth in this Declaration, without distinction of any kind, such as race, colour, sex, language, religion, political or other opinion, national or social origin, property, birth or other status. Furthermore, no distinction shall be made on the basis of the political, jurisdictional or international status of the country or territory to which a person belongs, whether it be independent, trust, non-self-governing or under any other limitation of sovereignty."

82 Art. 2, para. 1 of the ICCPR: "Each State Party to the present Covenant undertakes to respect and to ensure to all individuals within its territory and subject to its jurisdiction the rights recognized in the present Covenant, without distinction of any kind, such as race, colour, sex, language, religion, political or other opinion, national or social origin, property, birth or other status."

83 Art. 2, para. 2 of the ICESCR: "The States Parties to the present Covenant undertake to guarantee that the rights enunciated in the present Covenant will be exercised without discrimination of any kind as to race, colour, sex, language, religion, political or other opinion, national or social origin, property, birth or other status."

84 See HRI/GEN/1/Rev.9, November 10, 1989, para. 1 *et subseq.*

While the wording of the Peninsula Principles' provision differs from the usual nondiscrimination prohibition, as it does not list the grounds of discrimination, the same standards apply, since this Principle seeks to put climate displaced persons on the same footing as all other persons within a country. The International Bill of Human Rights generally lists race, color, sex, language, religion, political or other opinion, national or social origin, property, birth or other status, which is not however seen as exhaustive as the *other status* suggests, which may include disability as well as age and even sexual orientation.[85] It is interesting to note that amidst the initial drafts of the Peninsula Principles, tenet 3 had a different sequence in the title "Rights, Freedoms and Non-Discrimination" and contained a catalogue of grounds of discrimination such as age, legal, social or marital status, as well as sexual orientation.[86] Finally, it was decided to include an open nondiscriminatory clause tying it with the corpus of international and human rights law for reasons of efficiency and clarity and to rather emphasize the displacement setting, suggesting the repeal of unjust or arbitrary laws and any other direct and indirect discrimination of climate displaced persons.

It is one of the specificities of Peninsula Principle 3 that it explicitly claims affirmative action by the State to repeal any type of discriminatory legislation. This will require a thorough review of applicable law, which can be a tedious task and moreover requires expert knowledge of the local circumstances and specificities including the local language and legal system in place.

A provision or law may itself be discriminatory or it could be applied in a discriminatory manner. A particularly relevant example for climate displaced persons are so-called abandonment laws, which may be applied for the purpose of appropriation of property of those persons who had to flee the adverse effects of climate change and whose return to their properties had not been possible within the statutory timelines set by theses laws.[87] Repeal could be partial, focusing on individual provisions, or may concern a specific law as a whole. It can occur with or without reenactment and even through derogation as *lex posterior derogat legi priori*

85 CCPR/C/35/D/196/1985, April 3, 1989, paras. 9.4, 9.5 and 10.

86 Amidst the initial drafts (on file with the author) of the Peninsula Principles on Climate Displacement within States, an earlier wording of Principle 3 read as follows: "Principle 3: Rights, freedoms and non-discrimination (a) States shall not discriminate against climate displaced persons on the basis of their displacement or race; colour; gender; sexual orientation; gender identity; language; religion; belief; political or other opinion; national, ethnic or social origin; legal or social status; caste; age; disability; property ownership; place of birth; or any other status. (b) Climate displaced persons experiencing climate displacement shall enjoy, in full equality, the same rights and freedoms under international and domestic law as do other persons in their country, in particular housing, land and property rights."

87 Regarding the discriminatory application of abandonment laws, see, e.g., E/CN.4/2006/41/Add.2, March 21, 2006; UNHCR and Sri Lanka Human Rights Commission, Land, Housing and Property: Proposals to the Parties for Comprehensively Addressing Land, Housing and Property Rights in the Context of Refugee and IDP Return within and to Sri Lanka, 2003; R. C. Williams, Property, 2010, p. 379.

(i.e. that a later law repeals an earlier). This particular prominence placed on prohibiting any normative, practical or juridical bias on the basis of potential or actual displacement alone or in conjunction with other grounds follows also from negative experiences with regard to discriminatory laws and practices enhancing and fortifying displacement in post-conflict settings, and particularly ethnic cleansing scenarios, where so-called absentee laws facilitated the acquisition of HLP rights of displaced persons and refugees who had to flee and could not stay and reaffirm their rights locally on-site.[88] The UN Principles on Housing and Property Restitution for Refugees and Displaced Persons (i.e. the Pinheiro Principles) helped to counter the tendency – certainly out of the layman sphere, but still very much existent – that displaced persons' rights, and especially their HLP rights, vanish or that the concerned person had renounced them by virtue of being just displaced.

The Peninsula Principles' scope with regard to nondiscrimination is not limited to post-disaster vulnerability and needs assessments. On the contrary, through the emphasis on planning, the application of the Principles suggests that pre-disaster assessments of inequalities and vulnerabilities have to adopt a methodology apt to identify notably patterns of discrimination of any kind from race, color, sex, language, religion or belief, political or other opinion, national, ethnic or social origin, legal or social status, age, disability, property, birth, or on any other similar criteria.[89]

This draws from real-case scenarios as for example in the post-tsunami period certain governments exploited adaptation strategy plans and policies.[90] The authorities claimed emergency protection requirements and needs and created on this basis so-called coastal buffer zones, which prevented low-income persons affected by displacement from returning to their areas of origin, and obliged whole communities in coastal areas to relocate in an unplanned and, hence, unprepared manner. As a result, livelihoods were disrupted if not lost, creating or aggravating social tensions. For instance, fishing communities remained displaced and were permanently prevented from returning to their homes. In addition, exploitation and abuse of the post-tsunami situation occurred. Buffer zones were repurposed or rededicated to higher-income residential, commercial or industrial use, which at times, through the housing, land and properties vacated by the displaced, opened up opportunities for the travel and tourism businesses or for real estate speculators to extend their activities.[91] Similarly, postponement and interruptions in return processes as a consequence of well-established and deep-rooted arrays of discrimination,

88 FAO/IDMC/OCHA/OHCHR/UN-Habitat/UNHCR, Handbook, Housing and Property Restitution for Refugees and Displaced Persons, 2007, p. 88; Brookings Institution, Protecting Internally Displaced Persons: A Manual for Law and Policymakers, 2008, p. 174 *et seq.*

89 E/CN.4/1998/53/Add.2, Principle 4.

90 See, e.g., S. Leckie, An Integrated Approach to Climate Change Demands That Human Rights and Adaptation Strategies Are Pursued Hand-in-Hand, 2008, p. 18; D. Fisher, Domestic Regulation of International Humanitarian Relief in Disasters and Armed Conflict: A Comparative Analysis, 2007, p. 345 *et seq.*

91 *Cf.* A/66/270, para. 39 *et seq.*

disregard and inattention in Haiti lead to situations where certain displaced communities that benefited from some more favorable treatment by staying in the relatively better camps discriminated against others, which is extremely sensitive and undermines social cohesion toward a thriving community living together. This demonstrates the imperative and urgent need for a comprehensive approach focusing on communities rather than individuals, which the Peninsula Principles fully embrace.[92]

## VII PREVENTION COMMANDING RELOCATION

The human rights-based approach is a prerequisite for the adequacy of the response to climate displacement, namely prevention and remedy through planned relocation.[93] It was only in 2010, with the CAF, that migration, displacement and planned relocation were recognized as forms of adaptation to climate change.[94] Shaping adaptation measures in a manner that they prevent displacement will have priority, yet if these attempts fail, planned relocation will be the preferred option. In spite of this, an undercurrent of resistance to considering issues of planned relocation appears to exist, probably out of a concern that supporting such undertakings could weaken incentives and lessen political pressure on States and others, especially international governmental and nongovernmental stakeholders, to implement climate change mitigation measures that would in turn allow for communities and people to remain in their homes.[95]

Section II of the Peninsula Principles addresses the pre-displacement stage, and provides *stage directions* for all stakeholders in a consolidated manner on the steps that have to be taken in assessing and managing the climate displacement risk, the preparation and the implementation of planned relocations, including the identification of land, and the institutional design, as well as criteria for informed consent by and for the participation of climate displaced persons. The human rights imperative that underpins the Peninsula Principles seeks to ensure a transparent and coherent *modus operandi* in a multi-stakeholder approach.

The relocation aspect of the Peninsula Principles constituted the focus of discussions at the San Remo Consultation, which was held on March 12–14, 2014, organized by the UNHCR in cooperation with the Brookings Institute

92 *Ibid.*, para. 53 *et seq.*

93 See, for instance, S. Leckie, Land Solutions for Climate Displacement, 2014; S. Leckie, Finding Land Solutions to Climate Displacement: A Challenge Like Few Others, 2013; UNHCR, Planned Relocation, Disasters and Climate Change: Consolidating Good Practices and Preparing for the Future: Report, 2014, see also B. G. Ramcharan, Preventive Human Rights Strategies, p. 27 *et seq.*

94 E/CN.4/1998/53/Add.2; see also R. Cohen, Lessons from the Development of the Guiding Principles on Internal Displacement, 2014, p. 12 *et seq.*

95 UNHCR, Planned Relocation, Disasters and Climate Change: Consolidating Good Practices and Preparing for the Future, Background Document, 2014; L. Schipper, Conceptual History of Adaptation in the UNFCCC Process, 2006, p. 82 *et seq.*

and the Georgetown University. It was emphasized in this context that planned relocations as a solution are not a forgone conclusion, but rather represent the prudent exercise of responsibility.[96] The human rights proviso for planned relocations, as enshrined in the Peninsula Principles, and the need to draw on good practices from other efforts were the main points of consensual agreement.[97]

Similarly, the UN Advisory Committee (i.e. the think tank of the Human Rights Council) highlighted in its reflection paper and research proposal that "there is a pressing need to develop the national and international normative, institutional and implementation frameworks."[98]

### 1 *Nonpareils of the Peninsula Principles*

Tenets 9, 11 and 13 build the triumvirate of parameters for preventive and planned action. They span from the assessment phase to detect climate displacement risks, to the assertion of alternative lands for the climate displaced and finally the administrative–institutional frameworks for related assistance and provision of protection. The normative structure of the Peninsula Principles indicates that Principle 5 on prevention and avoidance under Part I constitutes a guiding principle, while Part II, Climate Displacement Preparation and Planning, propounds the concrete forms and basic elements of the preventive approach.

#### a Risk Assessment

The prerequisites for a planned relocation are defined in Principle 9, on climate displacement risk management, which constitutes the starting point of climate displacement action. The Principle lists the concrete actions States should take "in terms of climate displacement risk management, monitoring and modeling, using a rights-based approach." The detailed handling instruction for governments and State authorities includes that they take measures to:

a. identify, design and implement risk management strategies, including risk reduction, risk transfer and risk sharing mechanisms, in relation to climate displacement;

96 UNHCR, Planned Relocation, Disasters and Climate Change: Consolidating Good Practices and Preparing for the Future, Background Document, 2014.

97 *Ibid.*

98 A/HRC/AC/17/2, annex IV, para. 41; the Advisory Committee proposed in this regard to prepare guidelines on climate displacement and human rights, based on existing frameworks such as the Guiding Principles on IDPs and the Peninsula Principles on Climate Displacement within States, or to prepare a report on the question of a human rights-based planned relocation to cases of looming climate displacement. *Ibid.*, para. 42.

b. undertake systematic observation and monitoring of, and disaggregated data collection at the household, local, regional and national levels on, current and anticipated climate displacement;
c. enhance sharing, access to and the use of such data at the household, local, regional and national levels, mindful of the need for data protection and predetermined use of data, facilitate the assessment and management of climate displacement;
d. model likely climate displacement scenarios (including timeframes and financial implications), locations threatened by climate change, and possible relocation sites for climate displaced persons;
e. integrate relocation rights, procedures and mechanisms, as defined in these Peninsula Principles, within national laws and policies.

The degree of concreteness and detail of Principle 9 stems from the fact that it could actually draw on the good practices that the work of the practitioners with affected communities elicited and that informed the Principles' drafting process. It was also a subject of the discussion in the context of the *travaux préparatoires*. Some members voiced concerns about the level of detail, which would be too prescriptive and impede the handling of relocations. They argued that too many conditions and requirements could ultimately obstruct implementation. The competences for government action are rarely centralized, hence imaging that the authorities at various levels (i.e. central, regional and local) would simply issue decisions and take action on climate displacement risk management and analysis seemed illusionary to some. Others, however, argued that a detailed listing of matters to address and a minimum of sequencing and regulatory guidance would ensure that approaches are not flawed. They held that this innovative aspect of the Principles needed to be detailed to facilitate the application of the Peninsula Principles and to somehow demystify rights-based planned relocations.[99] This level of detail was also warranted given the experience with involuntary development-based resettlements, which rendered those affected more vulnerable as a result of the loss of resources and livelihoods.[100]

There was consensus about the role and responsibility of climate displaced persons throughout the process, including in the risk assessment to monitor, report and document climatic changes and sociocultural impacts. These records of information and accounts of communities and persons, potentially threatened by climate displacement, should obviously be conducted in cooperation with competent central, regional and in particular local authorities and entrusted actors and stakeholders. The systematic observation and monitoring of climate displacement risks and the disaggregated data collection at the household level are meant to ensure that those most immediately affected have a participatory role in the gathering of data

99 Peninsula Principles' *Travaux Préparatoires*, August 2013 – on file with the author.
100 J. Barnett, M. Webber, Migration as Adaptation: Opportunities and Limits, 2012, p. 53.

during the risk assessment process and that it draws from good practices as advanced by Bradley May and Ryan Plummer or Robin Bronen.[101]

The principle of subsidiarity manifests itself in this context. Methodologies and processes for rights-based participatory consultations and fact-finding that would inform decision-making are deemed essential components for good governance as described by Bronen and Gudmundur Alfredsson for instance.[102] The determination of climate displacement risks and whether relocation is required can only be taken sustainably, if local data collection feeds into the process and with the participatory collaboration of the concerned party – that is those *facing* climate displacement.

The difficulty in the instant case however lies in the absence of clear-cut measurable criteria of general validity to determine when relocation has to occur. The use of local data, comparative analysis and participatory consultation along with assessments, monitoring and evaluation are instruments intended to assist the governmental, administrative decision-making processes.[103]

Continuous monitoring of climatic changes and their impact on individuals and communities is not just an account of climatic developments, but a prerequisite for resilience measures and ultimately relocation. Prognostics need to take into account that climatic changes are dynamic and may differ throughout temporal and spatial scales; and vary along economic, social, geographic, demographic, cultural, institutional, governance and environmental factors.[104] Information, studies, data and assessments as well as the resulting prognostics at the global or regional level are most relevant in assessing the risk at the local level. Information exchange and interactive communication is hence vital to determining climate displacement menaces.[105]

Of note is also the guidance that Peninsula Principle 9(f) provides for the development of institutional frameworks, procedures and mechanisms with the participation

101 B. May, R. Plummer, Accommodating the Challenges of Climate Change Adaptation and Governance in Conventional Risk Management: Adaptive Collaborative Risk Management (ACRM), 2011; R. Bronen, Forced Migration of Alaskan Indigenous Communities Due to Climate Change, 2010; R. Bronen, Climate-Induced Community Relocations: Creating an Adaptive Governance Framework Based in Human Rights Doctrine, 2011, p. 101 *et seq.*; R. Bronen, S. F. Chapin, Adaptive Governance and Institutional Strategies for Climate-Induced Community Relocations in Alaska, 2013; R. Bronen, Climate Displacement: Preparation and Planning, 2016, p. 89 *et seq.*; R. Bronen, Choice and Necessity: Relocations in the Arctic and South Pacific, 2014, p. 17 *et seq.*

102 R. Bronen, Climate-Induced Community Relocations: Creating an Adaptive Governance Framework Based in Human Rights Doctrine, 2011, p. 101 *et seq.*; R. Bronen, S. F. Chapin, Adaptive Governance and Institutional Strategies for Climate-Induced Community Relocations in Alaska, 2013; R. Bronen, Climate Displacement: Preparation and Planning, 2016, p. 89 *et seq.*; G. Alfredsson, Good Governance in the Arctic, 2013.

103 *Cf.* Brookings Institution, From Responsibility to Response: Assessing National Approaches to Internal Displacement, p. 113 *et seq.*

104 IPCC, Managing the Risks of Extreme Events and Disasters to Advance Climate Change Adaptation, Summary for Policymakers, 2012, p. 3 *et seq.*

105 D. A. Lewis, The Relocation of Development from Coastal Hazards through Publicly Funded Acquisition Programs: Examples and Lessons from the Gulf Coast, 2012, p. 98 *et seq.*

of individuals, households and communities. It outlines the following course of action:

(i) identify indicators that will, with as much precision as possible, classify where, at what point in time, and relevant to whom, relocation will be required as a means of providing durable solutions to those affected;
(ii) require and facilitate governmental technical assistance and funding; and
(iii) outline steps individuals, households and communities can take prior to climate displacement in order to receive such technical assistance and financial support.

Following from this, a socioeconomic situational analysis and displacement forecast is required to determine which communities will most probably need to relocate. Bronen suggests in this regard social–ecological indicators, which need to be specific to ecosystems, geographical regions, and social, political and economic systems. Proposed indicators of socioecological vulnerability include the repeated loss of community infrastructure; the imminent danger to the community from ongoing ecological changes and repeated random extreme weather events; no ability for community expansion; numbers of evacuation incidents and numbers of people evacuated; predicted rates of environmental change (e.g. sea-level rise) from the IPCC; repeated failure of hazard mitigation measures; a lack of viable access to transport, potable water, communication systems, power and waste disposal; and decline in socioeconomic indicators, including food security, loss of livelihood and public health. According to Bronen these indicators are yardsticks to determine whether and, if so, when relocation is necessary to protect the safety of people and communities.[106]

### b Land Identification

The second pillar of the climate displacement prevention triumvirate is Principle 11, on land identification, habitability and use. In the event of impossibility of return, land is the solution; hence the relocation site is a vital aspect of displacement prevention. Principle 11 consequently emphasizes the States' responsibility based on the assumption that it *disposes* over sufficient adequate land for those threatened by climate displacement. At this stage, meticulous assessments and preparations are required to shape the responses to the specific context, including such issues as spatial or building and construction planning.

Land solutions imply moreover the provision of livelihoods. This is at times complex, given the spectrum of issues that can arise in relation to HLP rights in combination with subsistence and livelihoods imperatives for climate displaced persons. In addition to challenges stemming from property determination and

[106] R. Bronen, Climate-Induced Community Relocations: Creating an Adaptive Governance Framework Based in Human Rights Doctrine, 2011, p. 101 *et seq.*; R. Bronen, Climate Displacement: Preparation and Planning, 2016, p. 89 *et seq.*

protection measures in conjunction with legal pluralism, spatial and habitation issues need adequate consideration at the earliest point in the planning phase so as to be able to qualify potential relocation sites as habitable and usable. Moreover, there might be instances where alternative land would be vacant but would require conversion to agricultural land and where this would conflict with public interest, for instance if it entails changing the land use of protected areas or nature reserves. In view of these intricacies, Principle 11(a)(i) requires States to "identify, acquire and reserve sufficient, suitable, habitable and appropriate public and other land to provide viable and affordable land-based solutions to climate displacement, including through a National Climate Land Bank." The actionable recommendation to create a National Climate Land Bank inscribes into the set of preventive measures seeking to prompt governments to enact relevant policies and laws in anticipation of climate displacement and for future planned relocation. Ensuring that HLP rights protection is provided for, procedurally and substantively, is indispensable for a legally sound, coherent and practical framework.

A land bank could assist State authorities in the provision of land and livelihoods, – as restitution in lieu of the loss resulting from climate displacement – and to expedite relocation to pre-identified land. Furthermore, it is necessary to establish geographically pertinent consistent standards for the assessment of the feasibility and habitability of land for planned relocation. In this context, Principle 11 does not spell out detailed requirements. While this falls within the ambit of domestic spatial as well as building and construction law and regulations, the Principles lack specific guidance in this crucial area for legislators. Habitability and feasibility as broader concepts provide nevertheless a sound benchmark for human rights compliance of national frameworks.

Moreover, Principle 11(d) asks State authorities to embrace the idea of establishing dedicated institutional mechanisms to provide for effective HLP rights protection and thus to ensure that livelihood rights can actually be realized. This is very important in the event of informal or customary HLP rights, occupancy rights, customary usage rights or other rights in rem. A particular mention of assurances that such rights to access traditional lands and waters – for example, for hunting, grazing, fishing and religious purposes – are maintained or similarly replicated seems necessary, given that the decision-making competencies with regard to public property rights and construction and spatial planning law are often scattered at different governmental levels.

### c Institutional Frameworks

Principle 13, the third prevention pillar, is dedicated to the institutional frameworks to support and facilitate the provision of assistance and protection. The guidance does not substitute itself to the required development of an appropriate framework that would obviously need to be tailored to the situation and ideally be based on a participatory consultation prior to its establishment. The provision starts with urging States' commitment to "strengthen national capacities and capabilities to identify and address the protection and assistance needs of climate displaced persons through the establishment of effective institutional frameworks and the inclusion of climate displacement in National Adaptation Programmes of Action as appropriate." The Principle also calls for a coherent and concerted approach of all responsible and competent authorities and urges sufficient resourcing. In the case of lack of necessary funding, this entails that technical assistance and capacity building *et al.* be provided. Principle 13 tries to ensure cross-departmental commitment to conduct and adequately plan climate-induced relocation. It also seeks to instill a focus shift from emergency response to prevention and for an enhanced focus on building climate resilience of communities.[107]

The centrality of resilience has been reiterated at a meeting on March 18, 2015, on the margins of the 28th session of the UN Human Rights Council on "Building Climate Resilience: The rights of groups in focus." This event brought together several UN Special Rapporteurs of the Human Rights Council, such as on the human rights of internally displaced persons, the full enjoyment of all human rights by older persons, on the rights of persons with disabilities and on the rights of indigenous peoples, as well as governments, key humanitarian actors such as the OCHA and the UNHCR, the academia and NGOs.

The purpose of this event was to create awareness and rally key actors behind the need for planning as a key feature of prevention for climate displacement and disaster risk reduction. Margareta Wahlström,[108] the then head of the UNISDR (i.e. the focal point in the United Nations system for the coordination of disaster reduction), was scheduled to participate in the meeting in Geneva, but had eventually to remain in Japan for the negotiations of the Sendai Framework. Although she could not actively participate, she supported the event and carried the message from Geneva.[109]

Under her leadership and in her capacity as Special Representative of the Secretary-General in charge of supporting the implementation of the Hyogo Framework for Action 2005–2015: Building the Resilience of Nations and Communities to Disasters throughout the world and facilitating the process of developing a post-2015 framework for disaster risk reduction, the strong message from the meeting in Geneva permeated the negotiations in Sendai and ultimately found its way into the new framework for Disaster Risk Reduction 2015–2030.[110]

107 *Cf.* J. Rodin, The Resilience Dividend: Managing Disruption, Avoiding Disaster, and Growing Stronger in an Unpredictable World, 2015, p. 159 *et subseq.*

108 https://unhabitat.org/ms-margareta-wahlstrom/ (accessed March 22, 2019).

109 See PR/OHCHR/20March2015, New framework for Disaster Risk Reduction.

110 The strong statement stressed that "[m]oving from words to effective actions requires the full and effective participation and accessibility of individuals and communities affected, including women, persons with disabilities, indigenous peoples, older persons, as well as those displaced internally notably by the effects of climate change, in all phases of disaster risk reduction and building resilience strategies and policies." The human rights experts drew special attention to the gender, age, disability and cultural perspectives included in the new framework for Disaster Risk Reduction 2015–2030, and described it as "a significant progress." "The concerns and rights of groups must be taken into consideration for a successful implementation of the Sendai Framework," they stressed. "All

Against the background of the terrible events in Vanuatu, where the destruction brought upon by Cyclone Pam left thousands of persons displaced or insecurely housed, and mindful of the Sendai negotiations, a group of UN experts called for the protection of human rights of *all* in disaster response.[111] The imperative of founding recovery and rebuilding efforts on the human rights framework penetrated the sphere of negotiations and the outcome framework itself, which affirms that "[t]here has to be a broader and a more people-centred preventive approach to disaster risk."[112]

Again, this *focus shift* to rights-based prevention and planning was reconfirmed at the first World Humanitarian Summit, which was held on May 23–24, 2016, in Istanbul, Turkey. It brought together some 9,000 participants from 173 Member States, including 55 heads of state, thousands of representatives from civil society and NGOs, and hundreds of private sector representatives, with the shared objective of better preventing and responding to humanitarian crises.[113] It was the culmination of a process of consultations and provided a unique platform to share and discuss ideas, practices and solutions. It sought to achieve high-level commitments at the national, regional and international levels to inspire an ongoing process of change and improvement in responses to crises.[114] Again, the ostensible quest for practical assistance to States and other stakeholders was a key demand. The Special Rapporteur on the human rights of internally displaced persons, Chaloka Beyani,[115] who was a key participant in the World Humanitarian Summit, recommended the expansion of the scope and nature of collaboration and assistance beyond the immediate aftermath of disaster. He urged that "prevention, preparedness and longer-term recovery should be part of a new global compact" to be able to confront the challenges of conflicts, disasters and climate change.

In line with the increasing international *focus* on risk reduction and resilience in the 2015 Sendai Framework and the 2015 Paris Agreement on climate change, the World Humanitarian Summit addressed displacement in the context of disasters and climate change, and participants called for the development of an international mechanism and legal framework for the protection of those displaced by the adverse

individuals, without discrimination, must be considered as a resource for disaster risk reduction and resilience." See *ibid.*; *cf.* also A/CONF.224/CRP.1, Sendai Framework for Disaster Risk Reduction 2015–2030, March 18, 2015; IOM, Taking Sendai Forward, IOM Strategic Work Plan on Disaster Risk Reduction & Resilience 2017–2020, 2013.

111 PR/OHCHR/20March2015, New framework for Disaster Risk Reduction; PR/OHCHR/20March2015, Vanuatu.

112 Sendai Framework for Disaster Risk Reduction 2015–2030; the Framework was adopted at the Third UN World Conference in Sendai, Japan, on March 18, 2015. It is the outcome of stakeholder consultations initiated in March 2012 and intergovernmental negotiations from July 2014 to March 2015, supported by UNISDR at the request of the UN General Assembly.

113 See the Chair's summary, Standing up for humanity; A/71/279, para. 12.

114 A/71/279, para. 12 *et subseq.*

115 *Ibid.*, paras. 36, 84 *et seq.*, and para. 90.

impacts of climate change.[116] The Special Rapporteur on the human rights of internally displaced persons in this regard also noted the need to transcend the humanitarian–development divide, to allow for an integrated approach that invests in building self-reliance and resilience. Further to his participation in the Summit, he issued a powerful call for action: "[R]educing displacement as a result of climate change, sudden and slow-onset disasters and conflict, we must act now and establish effective safeguards and benchmarks that will avoid future lost generations of displaced. No one should be left behind and the World Humanitarian Summit has taken important steps, but its commitments and promise must be realized in practice."[117]

While this may not be a *novus habitus mentis*, it reflects the new mindset that reigns in the United Nations organization and the humanitarian sphere inscribing itself into the Secretary-General's vision on prevention as the guiding principle of international action.[118]

## 2 *Participatory Governance*

A prerequisite for sustainable and successful planned relocation undertakings is the partaking of affected people and communities. Therefore, the design of processes and the contribution of the relevant institutions in conducting people's engagement are crucial.[119]

116 *Ibid.*

117 *Ibid.*, para. 90.

118 See UN Secretary-General Priorities: Prevention

1. Support the development and implementation of national disaster risk reduction plans that address growing challenges of climate change, environmental degradation, urbanization and population growth. Special emphasis should be placed on the least and most vulnerable countries, including by providing a platform for South–South cooperation and facilitating the use of innovative methods and technologies.
2. Prioritize early warning and early action on preventing violent conflict by:
   - Mapping, linking, collecting and integrating information from across the international system;
   - Supporting national capacities for facilitation and dialogue;
   - Ensuring that UN good offices, mediation, crisis response and peacebuilding services are easily and rapidly deployable.
3. Advance a preventive approach to human rights by:
   - Developing a policy framework that identifies basic elements needed to prevent human rights violations;
   - Establishing a preventive matrix that will chart progress and gaps in the use of a range of human rights instruments;
   - Advancing the responsibility to protect agenda.
4. Build resilience to external economic and financial shocks by helping countries identify vulnerabilities rapidly and adopt adequate social safety nets and policies that promote job-led growth.

www.un.org/sg/en/priorities/prevention.shtml (accessed May 21, 2019).

119 See, e.g., E/C.16/2007/2; F. Fischer, Participatory Governance: From Theory to Practice, 2012, p. 457 *et subseq*.

Partaking and participation have to be understood as active involvement of affected communities and people as well as those with a legitimate interest in the formulation of regulatory objectives, policies and approaches, or in the drafting of regulatory texts governing in the instant case planned relocation undertakings of climate displaced persons. Thereby, the governance's overall objective should be to facilitate implementation and improve compliance with the applicable and relevant body of law, as well as to build consensus and generate political support. National or local authorities are likely to offer stakeholders a role in regulatory development, implementation and/or enforcement if government discernment exists, to increase the sense of ownership of and/or commitment to the regulations, guidelines and decisions, beyond what is probably realizable merely by means of a consultative approach.

Principle 10, on participation and consent, constitutes a *conditio sine qua non* for human rights-compliant planned relocations as a general rule. The consent requirement is stringent and can only be waived in case of "exceptional circumstances" (i.e. "when necessary to protect public health and safety or when individuals, households and communities face imminent loss of life or limb"). This reflects the regular general legal and polity situation and position.

*Consent* is the key term, expressing the prerequisite of effective participation regarding such relocation efforts. The requirement for full and informed consent is not limited to those who are to be relocated, but extends to the host communities as well. Principle 10 provides "that no relocation shall take place unless individuals, households and communities (both displaced and host) provide full and informed consent for such relocation." The principle of free, prior and informed consent is an established principle in international law, even though there exists no universally agreed definition. In the context of relocations, it is for example enshrined in Article 6 of the Convention of the International Labour Organization (ILO) on Indigenous and Tribal Peoples in Independent Countries (169/1989), which explicitly mentions the principle of free and informed consent in relation to the relocation of indigenous peoples from their land. The UN Declaration on the Rights of Indigenous Peoples also explicitly recognizes the principle of free, prior and informed consent.[120] Article 19 of the Declaration requires States to consult and cooperate in good faith with the indigenous peoples concerned through their own representative institutions in order to obtain their free, prior and informed consent before adopting and implementing legislative or administrative measures that may affect them. In the case of relocation of indigenous peoples from their lands or territories, there is even, according to Article 10 of the Declaration, an obligation by the State to obtain consent.

The Peninsula Principles encapsulate this emerging common understanding of the principle of free, prior and informed consent and provide a long checklist of measures to be taken by States, which includes the full and active participation

[120] A/RES/61/295 of September 13, 2007; see also E/CN.4/Sub.2/AC.4/2004/4.

of those affected in decision-making and implementation of decisions, to prepare a master relocation plan prior to any relocation and to ensure that, at all stages of the process, the rights of individuals, households and communities are protected.

The concrete form of the participatory process can be manifold; there are instances where progress toward implementing participatory governance is measured by national level indicators through mechanisms such as *effective* Regional Advisory Councils and related meeting forums and even providing for the possibility of ad hoc meetings.

The Principles emphasize the requirement to put in place grievances and appeals proceedings. Such proceedings validate the participatory process and, at the same time, such remedy procedures generate a useful side effect, since they serve also as a vehicle to identify all issues and interests as well as potential matters of dispute. The list of issues generated through a functioning and effective grievance mechanism by merely indexing the matters of dispute such as conflicting claims, demands and entitlements is a rather reliable one, and awareness of sensitivities and obstacles is primordial for the relocation to be sustainable. Clearly, the efforts devoted at this stage of the process are more than valuable and correlate with the level of smoothness in conducting the relocation process. The participatory process in relation to planned relocation of climate displaced communities and persons remains without alternative, as knowledge acquisition as well as acquiescence and acceptance of the local populace and communities is essential for such relocation processes to succeed.[121]

## 3 *Adequate Consultation*

Adequate consultation can be facilitated and achieved by a constructive, participatory interactive dialogue. Such a process also builds the confidence of all stakeholders, and should have a certain precluding character, as those views that were not voiced within the framework of an all-inclusive dialogue may not, or only in exceptional circumstances, be taken into consideration at a later point. As a general model, what can be said about consultative processes is that they ought to encompass inner and outer spheres (i.e. the internal consensus building and the external consultation).[122] Both can consist of formal and informal components. In

121 See also the Aarhus Convention, i.e., the UN Convention on Access to Information, Public Participation in Decision-Making and Access to Justice in Environmental Matters, 1998; S. F. Martin, Managing Environmentally Induced Migration, 2009, p. 371 *et seq.*; R. Few, K. Brown, E. L. Tompkins, Public Participation and Climate Change Adaptation: Avoiding the Illusion of Inclusion, 2007; see also, B. Lim, E. Spanger-Siegfried, Adaptation Policy Frameworks for Climate Change: Developing Strategies, Policies and Measures, 2004; J. Barnett, M. Webber, Migration as Adaptation: Opportunities and Limits, 2012, p. 50; Displacement Solutions, Regulatory Obstacles to Rapid and Equitable Emergency and Interim Shelter Solutions after Natural Disasters, 2011, p. 31 *et subseq.*

122 See Brookings Institution, Moving beyond Rhetoric: Consultation and Participation with Populations Displaced by Conflict or Natural Disasters, 2008, p. 6 *et seq.*

addition to public meetings, interested parties should be provided with an opportunity to express their views in special meetings, popular discussions and other public fora. It is important that interested parties are enabled to intervene at an early stage of the process to assess the prospect for consensus building and to allow for necessary adjustments prior to implementation. Mixed committees of domestic authorities, representatives of international organizations, civil society actors and experts can also be very helpful in this regard. It should nevertheless be borne in mind that the emerging consensus that may be identified by means of popular consultations will often be a fragile one, which does not necessarily allow all the details of complex matters to be discussed, as this would expose the underlying and unresolved differences of opinion and conflicts. Some matters may have to be "black boxed" until a new societal consensus emerges. The overall value of consultative processes for decision-making, as well as for the sustainability of the outcome in terms of acceptance, abidance and implementation, cannot be emphasized enough. While the costs associated with such processes, particularly as a result of their time-intensiveness, is seemingly high, a rights-based process cannot be reduced to a simple cost–benefit analysis since it is essential for a sustainable and apprehended solution.[123] *Nihil de nobis, sine nobis*[124] is the overarching guiding principle in this regard.

The Aarhus Convention offers an excellent example in terms of incorporating procedural rights into human rights law. Public participation is a key element of the Convention, which in Principle 10 outlines that "[e]nvironmental issues are best handled with the participation of all concerned citizens, at the relevant level."[125] Systematic and active information gathering and sharing are intrinsic to a transparent participatory approach, based on exchange and implicating the affected communities in the course of decision-making and subsequent action, given that those affected by disasters require to be kept abreast about the latest risk assessments and the obtainability of aid resources and other assistance.[126] This seeks to ensure that each individual has "appropriate access to information concerning the environment that is held by public authorities, including information on hazardous materials and activities in their communities, and the opportunity to participate in decisionmaking processes."[127] The Convention furthermore requires States to ensure "[e]ffective access to judicial and administrative proceedings, including redress and remedy."[128]

Consultations need moreover to be meaningful.[129] This means that it is not sufficient for a community facing climate displacement to be able to voice their

[123] See, e.g., the popular consultations in the Sudan, J. Gluck, Why Sudan's Popular Consultation Matters, 2010.

[124] i.e. Nothing about us, without us.

[125] See, e.g., Aarhus Convention, June 25, 1998.

[126] *Ibid.*

[127] *Ibid.*, Principle 10.

[128] *Ibid.*, Principle 10.

[129] *Cf.* Brookings Institution, Moving beyond Rhetoric: Consultation and Participation with Populations Displaced by Conflict or Natural Disasters, 2008, p. 30 *et seq.*

concerns, but that they need to be enabled, through the provision of adequate information, to ascertain the impact of the measures intended.[130] In the context of forced evictions, the Committee on Economic, Social and Cultural Rights (CESCR) even refers to the need of *genuine* consultations.[131]

## VIII PROTECTION AND ASSISTANCE THROUGHOUT THE DISPLACEMENT CYCLE

### 1 *Housing, Land and Livelihood*

Those who were relocated as well as those displaced by climate change require specific assistance, including in terms of provision of adequate housing, land and livelihood.

Peninsula Principle 15, which is concerned with the actual displacement stage, reiterates that States should respect, protect and fulfill the right to adequate housing of climate displaced persons experiencing displacement. It also urges, where climate displacement results in the inability of climate displaced persons to return to previous sources of livelihood, that appropriate measures be taken to ensure such livelihoods can be continued in a sustainable manner and, notably, will not result in further displacement.

Similarly, situations of planned relocations, as a deliberate and strategic decision, should certainly provide for an adequate infrastructure that suffices the right to an adequate standard of living, including adequate housing as one of its components. States and local authorities need to take appropriate measures that promote livelihoods, acquisition of new skills and economic prosperity for both those relocated and the host communities. Public utilities need to be functional in order to ensure basic services, adequate and affordable housing and education. Interestingly, the Peninsula Principles term an equity standard among relocated and the host community. It is important not to infringe the rights of the host community while attempting to protect the rights of climate displaced.

Discrimination and bias even by negligence have to be avoided, and equality needs to be ensured by upholding equal rights for all, the displaced persons as well as those who have not been relocated. Otherwise, there is a probability that the relocation effort may be undermined, if social, income or economic inequality occurs and prevails. Henceforth, it is necessary to concoct inventive social protection

130 *Maya Indigenous Community of the Toledo District* v. *Belize* [2004], IACHR Case 12.053, Report No. 40/04, OEA/Ser.L/V/II.122, Doc. 5, Rev.1, at 727, paras 154–155; ILO, Convention No. 169 Concerning Indigenous and Tribal Peoples; CCPR/C/52D/511/1992, November 8, 1994, para. 9.5; Social and Economic Rights Action Center and the Center for Economic and Social Rights ./. Nigeria ('SERAC ./. Nigeria – the Ogoniland Case'), AfCHPR, Communication 155/96 (2002), para. 53.

131 E/1998/22, para. 15.

mechanisms to limit the impacts of climate displacement and adopt a handling that safeguards social sustainability.

The right to adequate housing is enshrined in Article 25(1) of the Universal Declaration of Human Rights as part of the right to a standard of living adequate for the health and well-being of the individual and of their family. It is also included in Article 11(1) of the International Covenant on Economic, Social and Cultural Rights. Building on this, Principle 18 of the Guiding Principles provides that "[a]ll internally displaced persons have the right to an adequate standard of living" and "[a]t the minimum, regardless of the circumstances, and without discrimination, competent authorities shall provide internally displaced persons with and ensure safe access to: [. . .] basic shelter and housing."

In its General Comment No. 4 on the right to adequate housing, the CESCR defined the *adequacy* as encompassing security of tenure, availability of services, materials, facilities and infrastructure, affordability, habitability, accessibility, location and cultural adequacy.[132]

Another aspect that is particularly key in the context of planned relocations is that relocation plans need to contain measures to compensate climate displaced persons for lost HLP (see Peninsula Principle 11(d)). The focus here is not on facilitating the return of climate displaced persons, but that their rights to housing, land and properties are protected while they have to relocate to other areas. It is an acknowledgment that their rights continue to exist in spite of the fact that they have to leave and that therefore they are entitled to some form of compensation. Principle 11(d)(ii) also specifically requests States to make "assurances that housing, land, property and livelihood rights will be met for all climate displaced persons." It further specifies which rights would fall within the meaning of HLP rights. It includes not only formal rights, but also informal land rights, customary land rights, occupancy rights or rights of customary usage, and there is a need for assurances that all these rights will continue to exist and be protected.

In terms of designing compensation mechanisms, States can draw on the Pinheiro Principles, which informed the drafting of the Peninsula Principles on these particular points. Principle 2.1 of the Pinheiro Principles provides that "[a]ll [. . .] displaced persons have the right [. . .] to be compensated for any housing, land and/or property that is factually impossible to restore as determined by an independent, impartial tribunal." The Pinheiro Principles, even though their rationale has been, first and foremost, the facilitation of return of displaced persons, provide very practical guidance on how to deal with HLP restitution, including the institutional design, restitution and compensation as an ultima ratio. While compensation can be in cash or in kind (i.e. nonpecuniary), it always has to be adequate, just and equitable.

[132] E/1992/23.

### 2 *Budget and Funding*

Sound financial resources and an adaptation funding play a central role in addressing climate displacement. The objective in developing corresponding budgets is coherence, as they need to enable policy and polity action on climate displacement. Thus, earmarking budgetary provisions to effectively resource local institutions and mechanisms dealing with climate displacement is essential. Budget analysis, mapping of the relevant actors and of their respective budgetary competencies are certainly complex matters, as this requires a broad approach unifying local, regional and national levels toward a consolidated budget plan for climate displacement action. The different State authorities need to be committed to provide budgetary allocations and earmarking for climate displacement management. Also indispensable is an economic impetus as well as a stimulus for investment in prevention and related risk reduction. Encouragement and tracing of such investments will give estimates of the expenses and return on investments on the grounds of evidenced information and data, which can then be further used.[133]

Accordingly, Peninsula Principle 7 emphasizes that it is not sufficient to establish institutions and mechanisms at all levels of government (local, regional and national) for their implementation, but that in order to give effect to its provisions "specially earmarked budgetary allocations and other resources to facilitate that implementation" are required. This is in line with States' obligations under international human rights law. It is not sufficient for States to legislate, but they also commit to make the necessary financial provisions in line with the progressive realization of rights.[134]

### 3 *On-site Mediation and Grievance Procedures*

Peninsula Principle 16 stipulates that "[c]limate displaced persons experiencing displacement but who have not yet been relocated and whose rights have been violated shall have fair and equitable access to appropriate remedies and compensation." This Principle acknowledges the fact that climate displaced persons are particularly vulnerable to violations of their rights, notably their HLP rights.[135]

Transparent and accessible redress procedures and proceedings are vital. Physical proximity, ensuring accessible and tangible redress mechanisms for climate displaced people are crucial in implementing the Peninsula Principles. Regardless of the respective displacement stage, it is imperative that mechanisms are in place at the local level providing effective remedies for mediation and grievance redress. Beyond these institutional mechanisms, individuals and communities also need to actually – *in realiter* – claim their rights, irrespective of their potential legal positions.[136]

[133] OECD, Integrating Human Rights into Development, 2013.

[134] E/1991/23; E/C.12/GC/20; E/C.12/2007/1, para. 3.

[135] E. Simperingham, The Responsibilities of States to Protect Climate-Displaced Persons, 2016, p. 128 *et subseq.*

[136] *Ibid.*

## IX RETURN OF CLIMATE DISPLACED PERSONS

The term *return* is intimately linked to the refugee notion. It designates the process by which a person who had been assigned refugee status returns to their country of origin. In the case of persons displaced internally due to the adverse effects of climate change, *return* refers to the place of habitual residence prior to the displacement. In analogy to the return of refugees, the returning climate displaced persons require specific assistance as well as protection during the transitional period until they have fully reintegrated into their societal environment and community.

The right to voluntary return in safety and dignity lies at the heart of Peninsula Principle 17. The core of the Principle is that return needs to be voluntary and effective, and it needs to be carried out in safety and with dignity. It is considered as one *durable solution* to displacement, alongside other options, i.e. *integration* at the location to which people were displaced, or *relocation* to another part of the country.

### 1 *Legal Foundation*

In spite of the return desideratum, for internally displaced persons this right is not explicit in existing international human rights treaties, even though the right of internally displaced persons to return to their original place of residence, or indeed to move to another safe place of their choice within their own country, has been asserted and reiterated in a myriad of international, regional and national jurisprudence, standards and normative deliberations.[137]

This norm draws on the right to freedom of movement and the right to choose one's residence, as embodied in Article 12 of the ICCPR. It can be subject to limitations, if properly reasoned and the necessary legal, judicial and temporal requirements are fulfilled in order to be justified – for instance for reasons of safety in the event of landslides or earthquakes.[138]

Guiding Principle 28(1) builds upon this and details that the competent authorities have the primary duty and responsibility to establish conditions, as well as provide the means that allow internally displaced persons to return voluntarily, in safety and with dignity to their homes or places of habitual residence, or to resettle voluntarily to another part of the country. Furthermore, authorities shall endeavor to facilitate the reintegration of returned or resettled internally displaced persons. The ILO's Convention No. 169 concerning Indigenous and Tribal Peoples also explicitly determines in Article 16(3) that "[w]henever possible, these peoples shall have the right to return to their traditional lands, as soon as the grounds for relocation cease to exist." In the event of impossibility of return, which requires reasoned assertion,

137 W. Kälin, Guiding Principles on Internal Displacement: Annotations, 2008, p. 127.

138 M. Nowak, UN Covenant on Civil and Political Rights: CCPR Commentary, 2005, Art. 12, para. 41; Brookings Institution, Protecting Internally Displaced Persons: A Manual for Law and Policymakers, 2008, p. 85 *et seq.*

those affected shall, pursuant to Article 16(4), "be provided in all possible cases with lands of quality and legal status at least equal to that of the lands previously occupied by them, suitable to provide for their present needs and future development."

In accordance with these State responsibilities and rights of internally displaced, the Peninsula Principles confer upon States an obligation to "develop a framework for the process of return in the event that displacement is temporary and return to homes, lands or places of habitual residence is possible and agreed to by those affected."

## 2 *Voluntariness of Return*

The prime choice – including for climate displaced persons – will *regularly* be return, if the essentials are given. Return is not a truism, but a rather complex undertaking since those affected do not depend on their volition but rather on circumstances (i.e. the ability, safety, possibility versus the impossibility in terms of HLP as well as livelihoods). A climate displaced person's decision to return home will normally coincide with improved circumstance (i.e. the threat or danger that had caused them to leave their place of habitual residence will have significantly diminished or disappeared). Another motivation could be of a comparative nature (i.e. that the conditions in the place of shelter have become worse than those presumably existing in the place of return or temporary location).

In accordance with Peninsula Principle 17(b), and in support of the axiom of voluntariness, "States should allow persons experiencing climate displacement to voluntarily return to their former homes, lands or places of habitual residence, and should facilitate their effective return in safety and with dignity, in circumstances where such homes, lands or places of habitual residence are habitable and where return does not pose significant risk to life or livelihood." The Principle recognizes that, in order "to exercise their right to freedom of movement and to choose their residence" and take the return decision, States have to provide displaced persons "with complete, objective, up-to-date and accurate information (including on physical, material and legal safety issues)."

In terms of the type of information to be provided to climate displaced persons, the IASC's Operational Guidelines on the Protection of Persons in Situations of Natural Disaster offer some clarity. The list includes security assessments of sites for return, local integration or settlement elsewhere in the country; conduct of comprehensive and accessible public information campaigns as well as grassroots communication strategies on return, local integration and settlement elsewhere in the country; establishing mechanisms such as media reports, databases, information centers, etc. to provide internally displaced persons with information on the conditions at the place of their former homes or locations identified for settlement elsewhere in the country, and organization of *go and see* visits.[139]

[139] IASC, Operational Guidelines on the Protection of Persons in Situations of Natural Disasters, 2011, p. 47 *et subseq.*; Brookings Institution, The Year That Shook the Rich: A Review of Natural Disasters

The voluntariness also implies a prohibition of forced return. Principle 15(d) of the Guiding Principles in this regards stipulates the explicit "right to be protected against forcible return to or resettlement in any place where their life, safety, liberty or health would be at risk." This prohibition of forced return is inspired by the principle of non-refoulement in international refugee and human rights law, which is enshrined in Article 33(1) of the 1951 Geneva Convention and is considered customary law and was translated to apply to situations of internal displacement.[140]

## 3 *Return in Safety and with Dignity*

The notion of safe return refers to a return that takes place under conditions of legal safety. The concept appears to be clearer in the context of political persecution and asylum. In this connection, return in safety relates to protection forms such as amnesties or public assurances of personal safety and physical integrity.

The German Basic Law for the Federal Republic of Germany starts its constitution with the irrevocable Article 1(1) on human dignity and stipulates: "Human dignity shall be inviolable. To respect and protect it shall be the duty of all state authority." Then, in Article 1(2) it reasserts that "[t]he German people therefore acknowledge inviolable and inalienable human rights as the basis of every community, of peace and of justice in the world." Finally, it determines that this eternal unchangeable guarantee binds all separated powers by virtue of stating in Article 1(3) that "[t]he following basic rights shall bind the legislature, the executive and the judiciary as directly applicable law."[141] The Austrian constitutional law for example does not use this expression explicitly, however a similar notion of dignity is established in the Austrian federal law through the immediate applicability of the European Convention on Human Rights and especially Articles 3 and 8,[142] and also derives from the normative content of the principle of equality.[143]

In practice, elements must include that climate displaced persons are not pressured; that they can return unconditionally and that if they are returning spontaneously they can do so at their own pace; that they are not arbitrarily separated from

in 2011, March 2012, p. 5 *et seq.*; Brookings Institution, On the Front Line of Climate Change and Displacement: Learning from and with Pacific Island Countries, 2011, p. 27 *et seq.*; for an illustration of the return issue, see also H. Kramer, V. Dzihic, Die Kosovo Bilanz: Scheitert die internationale Gemeinschaft, 2005, p. 42 *et seq.*

140 W. Kälin, Guiding Principles on Internal Displacement: Annotations, 2008, p. 69.

141 See Grundgesetz der Bundesrepublik Deutschland, Basic Law for the Federal Republic of Germany in the revised version published in the Federal Law Gazette Part III, classification number 100–1, as last amended by Art. 1 of the Act of December 23, 2014 (Federal Law Gazette I, p. 2438).

142 See, e.g., H. Schambeck, Der Begriff der „Natur der Sache": Ein Beitrag zur rechtsphilosophischen Grundlagenforschung, 1964, p. 53 *et subseq.*; S. Augeneder, The Denial of Human Dignity from a Legal and Philosophical Point of View, 2004, p. 193 *et subseq.*

143 See, e.g., F. Ermacora, Handbuch der Grundfreiheiten und Menschenrechte in Österreich, 1963, p. 39 *et seq.* and p. 59 *et subseq.*; K. Lachmayer, The Austrian Constitutional Court, 2017, p. 106 *et seq.*

family members; and that they are treated with respect and full acceptance by their national authorities, including the full restoration of their rights.[144]

The Peninsula Principles do not only specifically emphasize that States should facilitate the "effective return in safety and with dignity," but moreover provide some key elements to translate this dignified return into practice, requesting States to "provide transitional assistance to individuals, households and communities during the process of return until livelihoods and access to services are restored." These elements are also prerequisites for sustainable return and a durable solution.[145] The type of support will largely depend on the specific circumstances, but can be framed by virtue of the components of the right to an adequate standard of living as enshrined in the Universal Declaration on Human Rights, notably the right to adequate housing, water and sanitation, food and the right to health.[146] The efforts by the State itself will often have to be complemented by the international community through humanitarian assistance or within the framework of development cooperation.

Assisting internally displaced persons with returning to their places of origin requires precursory proceedings. Facts have to be ascertained as a precondition for finding an objective impossibility of return. It is not sufficient that only one party to the return proceedings claims the impossibility of return (e.g. a local authority declaring that it does not have the capacity to deliver basic services), unless this would affect the safety and health of those affected in a manner that would justify prolonged displacement. Similarly, subjective impossibility is insufficient to dismiss return quests as this would interfere with the rights of climate displaced to freedom of movement and choice of residence.[147] Displacement that was initially justified on grounds of safety becomes arbitrary if it continues longer than required.[148]

## X OPERATIONALIZING AND APPLYING THE PENINSULA PRINCIPLES

### 1 *Operational Dexterity*

Silo thinking and acting is ingrained in the conceded competence or mandate of many actors. The entrusted task and the discretionary power attached to it as well as the ownership to discharge the mandate might become an obstacle for global thinking and could thus impede effective coordination of domestic institutions.

144 UNHCR, Handbook Voluntary Repatriation: International Protection, 1996, para. 2.4.

145 Brookings Institution, When Displacement Ends: A Framework for Durable Solutions, 2007, p. 9 *et seq.*

146 *Ibid.*, p. 36 *et seq.*

147 S. Leckie, Returning Home: Housing and Property Restitution Rights of Refugees and Displaced Persons, 2003, p. 50 *et seq.*; K. Hassine, S. Leckie, The United Nations Principles on Housing and Property Restitution for Refugees and Displaced Persons: A Commentary, p. 156 *et seq.*; *cf.* BVerwG 7 B 93.02 (2003), para. 3; ECtHR, Demopoulos, para. 116; R. C. Williams, Introductory Note – European Court of Human Rights: Demopoulos v. Turkey, International Legal Materials 3/2010 (Oct. 2010).

148 *Cf.* E/CN.4/1998/53/Add.2, Guiding Principle 6.2.

Concerted action among various actors at the national level is a strategic challenge for the implementation of the Peninsula Principles.[149] Effectively implementing the Peninsula Principles entails overcoming the protectionist silo mentality at all levels, generating the necessary commitment to transgress institutional, thematic or organizational divisions at the domestic and the international levels. Climate displacement, including risk management, falls within the ambit of a number of ministries and local institutions, including the Ministry of Environmental Protection and Natural Resources, the Ministry of Agriculture, the Ministry of Finance, the Ministry of Justice and other relevant executive bodies concerned with urban/spatial planning. These should oversee the formulation of a long-term strategy on climate displacement in tandem with civil society through regular consultations. Civil society organizations and representatives of displaced and host communities should be involved in the strategy formulation in a structured and inclusive manner. Formal consultations at several stages of the process, including at the start and before finalization, are crucial for an informed and effective policy. In addition, the central government should work with national and international organizations and agencies, as well as with donor organizations, in order to obtain financial and informational support for the formulation and subsequent implementation of this long-term strategy.

## 2 *Emerging Practice*

### a Panama and the Case of Gunayala

Panama received a fact-finding visit conducted by Displacement Solutions from March 28 to April 9, 2014.[150] Displacement Solutions has since visited the region four times.

The objective of the initial visit was to study and examine the impact of rising sea levels on the displacement of communities in the autonomous region of Gunayala and the planned relocation of these communities from affected islands to the mainland. The mission used the Peninsula Principles as a guiding framework for assessing the situation and for making recommendations as to what action would be required to ensure that the human rights of communities affected by or threatened with climate displacement are fully respected and protected. Besides, the program of work also included human rights training and sensitization sessions of local communities and their leaders centered around the Peninsula Principles with the

149 *Cf.* Brookings Institution, From Responsibility to Response: Assessing National Approaches to Internal Displacement, p. 31 *et seq.* and p. 75 *et seq.*

150 *Cf.* Displacement Solutions, The Peninsula Principles in Action, Climate Change and Displacement in the Autonomous Region of Gunayala, Panama, Mission Report, 2014.

objective of empowering the affected as rights-holders. The idea was to enable them to take action themselves to claim their rights and to obtain the assistance they need and thus adopt an effective and sustainable approach to climate displacement.[151]

As to the initial visit's methodology and approach, the inclusion of international, regional and local experts is standard practice in exploring and assessing the situation in situ, as are meetings with the government and local authorities. Conducting it jointly with a renowned photojournalist as well as with grassroots organizations and *neighborhood committees* is outside the routine modus operandi,[152] but proved to be a resource-efficient way of impact maximization.[153] Documentation of the situation and awareness raising are crucial to generate support for preventive action. The idea was to avoid harm on the one hand and the usual mechanics of humanitarian and post-disaster assistance on the other hand.[154]

This fact-finding visit detected a potential for preventive action on the basis of the Peninsula Principles by proceeding with the planned relocation of the Gardi Sugdub community, the livelihoods of whom were (and still are) threatened by rising sea levels. Without such planned relocation, this community may at some point, when they will no longer be able to cope with the rising sea level, be forced to move and may end up in one of the slums of Panama City. This would not only put them in a situation of accrued vulnerability but also result in the loss of their livelihoods and traditional way of life. Another almost forty Guna communities will also eventually have to relocate to the mainland in the years to come.

The mission suggested that the Gardi Sugdub community should actively be involved in developing a credible and viable relocation plan and archetype course of action. An orderly and successful relocation of the Gardi Sugdub community could become

[151] See also, as a result of extensive consultations with climate displaced persons and their advocates directly, Displacement Solutions, The Rights of Climate Displaced Persons, A Quick Guide, 2015.

[152] Note that the Displacement Solutions visit was at the same time a follow-up visit for photojournalist Kadir van Lohuizen, who had visited Gunayala back in July 2011. Van Lohuizen works in collaboration with Displacement Solutions to document, through photography and video, how global warming is affecting people and communities worldwide. His work has also been published by the *New York Times* and other news outlets around the globe. See Displacement Solutions, The Peninsula Principles in Action, Climate Change and Displacement in the Autonomous Region of Gunayala, Panama, Mission Report, 2014, p. 8, and, e.g., www.nytimes.com/interactive/2014/03/27/world/climate-rising-seas.html?hp&_r=1 (accessed July 5, 2019).

[153] The Displacement Solutions mission was conducted by Carlos Arenas as substantive expert and photojournalist Kadir van Lohuizen in cooperation with Joana Abrego from the Centro de Incidencia Ambiental (CIAM), a leading Panamanian environmental organization. Guna General Congress (GGC), a grassroots organization, and the local district neighborhood committee as well as many individuals and actors of the local civil society supported the mission team. In Mandi Ubgigandub (Soledad Mandinga), the delegation met with the local authorities, ex-cacique Gilberto Arias. See Displacement Solutions, The Peninsula Principles in Action, Climate Change and Displacement in the Autonomous Region of Gunayala, Panama, Mission Report, 2014.

[154] Displacement Solutions, The Peninsula Principles in Action, Climate Change and Displacement in the Autonomous Region of Gunayala, Panama, Mission Report, 2014.

a model for the relocation of the other communities in the future. It could moreover become an international model for climate displaced people worldwide.[155]

The recommended course of action followed the tenets of the Peninsula Principles based on the finding that there was a need to acknowledge that rising sea levels are already causing climate displacement from the islands of Gunayala and to urgently initiate a process to prepare, plan and respond to climate displacement in the region.[156] It was moreover proposed that the Peninsula Principles be used as a guiding framework to establish institutional frameworks, procedures and mechanisms to address the problem. The authorities were asked to undertake immediate mitigation, adaptation and other preventative measures to give effect to the right of island communities threatened with climate displacement in Gunayala to remain in their homes and retain connections to the land on which they live for as long as possible.

It was also recommended to engage in a participatory planning process with the Guna communities and organizations in Gunayala to devise a master plan for relocation that addresses critical matters such as land acquisition, adequate housing, access to education, basic services and livelihoods to ensure the long-term sustainability of the process. Finally, the Displacement Solutions delegation suggested to initiate a process to identify land for relocation in recognition of the central role that land plays in any policy designed to address the consequences of climate displacement, and to also clearly earmark funds for acquiring land for relocation in Gunayala from national funds allocated for climate change adaptation measures. Furthermore, the need for capacity-building measures with the Guna people and their organizations was strongly encouraged. All of this should be tested in a pilot project for relocation from the island to the mainland.[157]

155 *Ibid.*, p. 32 *et seq.*

156 Of note, Scott Leckie's remarks on climate displacement and the reason for having a dedicated photojournalist: "We also learned during our fact-finding mission that the issue of climate displacement sometimes creates external false expectations or distorted truths. We were told, for instance, that from time to time some foreigner journalists come to Gunayala thinking that the islands where the Guna people live are permanently under water, and that they can get dramatic footage or pictures that literally show people living on flooded islands. We hope that the beautiful pictures from Kadir [van] Lohuizen that are included in this publication can help us see that the reality is a little more complex, especially considering that sea level rise is a certain but gradual event. Life goes on every day in Gunayala – but at the same time, little by little, the Gunas lose space in their fight with the sea. The magnitude and urgency of climate displacement is difficult to communicate. However, it is also clear that sensationalist approaches are never useful and are ethically questionable. As always, the truth speaks for itself, and people sympathetic to the fate of communities facing this problem worldwide can deal with the truth and still keep their interest and solidarity. We only hope that with the text and pictures of this report we have been able to provide some insights into the complexity of this issue in the magical and always surprising region of Gunayala." See Displacement Solutions, The Peninsula Principles in Action, Climate Change and Displacement in the Autonomous Region of Gunayala, Panama, Mission Report, 2014, p. 63.

157 *Ibid.*, p. 64 *et seq.*

This shows how the multi-perspective view of the Peninsula Principles provides a concrete basis for action and benchmarking to address climate displacement, including relocation. While the government made some initial commitments, and even offered in 2015 to build 300 houses for the relocation of Gardi Sugdub, and engaged in preliminary assessment of potential beneficiaries, little progress has been made. The challenges that remain are still significant despite this crucial groundwork undertaken by Displacement Solutions.

The rising acceptance of the Peninsula Principles is nevertheless exhibited by the fact that in this particular context they serves as *the* basis for benchmarking and the continued discussions in relation to the relocation process as well as its challenges and complexities in handling this threat of climate displacement.[158] Regretfully, the process has however stalled. Apart from some encouraging official statements, there have been few attempts to move toward a participatory planning process on relocation with the Gardi Sugdub community. This lack of implementation of commitments based on the Peninsula Principles shows that guidance cannot substitute deeds and the required political will.[159]

### b Colombia and the Case of Gramalote

The information that, on March 11, 2017, the Government of Colombia completed successfully a relocation exercise for a group of fifty-four families, moving them into the newly built city of Gramalote, is by itself an exemplary achievement. It took the Government approximately seven years after the destruction from the La Niña phenomena (i.e. extreme weather events in 2010 and 2011) to take concrete measures. The concerned group of people formed part of the approximately 3,300 climate displaced people affected by the massive mudslide that destroyed their community in December 2010.[160]

158 See Displacement Solutions, An Overview on the Relocation of Guna Indigenous Communities in Gunayala, Panama, Mission Report, 2016, p. 4 *et subseq.*; Displacement Solutions, The Peninsula Principles in Action, Climate Change and Displacement in the Autonomous Region of Gunayala, Panama, 2014; Displacement Solutions, One Step at a Time: The Relocation Process of the Gardi Sugdub Community in Gunayala, Panama, 2015; regarding the roundtable and Geneva discussions, see also Displacement Solutions, Land Access for Climate Displaced Persons and Communities, 2016.

159 See Displacement Solutions on the implementation status quo: "For the past three years, Displacement Solutions has been calling on the Panamanian government to apply the Peninsula Principles on Climate Displacement within States regarding the relocation of island indigenous communities in Gunayala, but until now this call has gone unanswered. The Panamanian government has made several promises to the community of Gardi Sugdub, but no concrete actions have been implemented yet."

160 Other families will continue moving throughout 2017 as their houses and other public buildings such as the school and hospital are finalized. It is expected that the 1,000 new houses will all be completed by March 2018. Despite challenges during the reconstruction process, especially during the initial stages, the outcomes so far are very encouraging. See Displacement Solutions, Climate Displacement and Planned Relocation in Colombia: The Case of Gramalote, 2015, p. 12 *et subseq.*

The Peninsula Principles were used as a blueprint for assessment, evaluation and monitoring of the situation, and thereby found their way into the related consultations and deliberations at the country level. As it is anticipated that the relocation process will continue for other families, monitoring and follow-up through the human rights lens of the Peninsula Principles should continue.

Whereas the Gramalote relocation process started before the Peninsula Principles actually came into existence, it allows to distill important lessons and to identify good practices. As the process continues to unfold, it remains to be seen whether Gramalote will set the example in translating the spirit of the Peninsula Principles, thereby adding to the global body of knowledge and experiences to be shared and exchanged.[161] For Displacement Solutions this is an important precedent:

> In many ways, Colombia is setting a positive example of how to deal fairly and equitably with effects of climate displacement and resultant planned relocation. In the immediate aftermath of the tragedy, the government of Colombia rightfully committed itself to rebuilding the entire town and ensuring that those affected would receive new homes for the homes they lost. Seven years later, it appears that the government is admirably keeping its promise. We hope governments everywhere will do the same and engage directly in preventing, repairing and resolving climate displacement in all of its forms, and in a manner that fully respects and protects the HLP and other rights of those affected.[162]

### c Strategic Choice: Bangladesh

Selecting Bangladesh as a programmatic priority country was a choice that suggested itself, as it is deemed one of the most vulnerable countries to climate displacement, and will have an estimated 20–30 million climate displaced persons in the coming decades.[163] Since 2011, Displacement Solutions has focused its efforts in Asia on the question of preventing and repairing climate displacement in Bangladesh due to the sheer scale of the challenge and related needs. Its activities are based on a strategic partnership with a leading Bangladeshi grassroots organization, YPSA.[164] Whereas each of the partners retains its specific role, there is a joint thematic approach. Responsibilities are shared along the organizations' complementarity and there is not – quite uniquely – a hierarchical *implementing partner* relationship between the global and the local organization. The rationale for this strategic partnership seemed

161 Displacement Solutions, Land Solutions for Climate Displacement Works in Colombia: Government-Supported Gramalote Relocation Moves Ahead, 2017.

162 See Displacement Solutions, Climate Displacement and Planned Relocation in Colombia: The Case of Gramalote, 2015, p. 12 *et subseq.*

163 For further information on the Bangladesh climate displacement initiative, see http://displacementsolutions.org/ds-initiatives/climate-change-and-displacement-initiative/bangladesh-climate-displacement/ (accessed March 2, 2019); Displacement Solutions, Building on Results: Climate Change and Displacement Initiative (CCDI), Funding Proposal 2014–2016, 2014, p. 49.

164 For an overview of YPSA's work, see www.ypsa.org (accessed April 20, 2019).

to be a decision to operate in the area with innovative solutions – planned relocation, land solutions and protection of HLP rights – that could generate information on good practices and any desiderata in the search for adequate responses in dealing with climate displacement. Thus, the partnership configuration relies on mutual exchanges with the objective to refine normative and regulatory responses. As a matter of fact, the groundwork in Khulna, Chittagong, Dhaka and elsewhere contributed significantly to the elaboration of the Peninsula Principles. It was precisely the cooperation with community-based organizations and YPSA that indicated a need to strengthen the existing normative framework on climate displacement and national legislation as it detected an important gap in this area. One priority that emerged was the protection of HLP rights of climate displaced persons in a rights-based manner so as to avoid that people spontaneously relocate to the slums of Dhaka or Chittagong and to hold out the prospects of their return to their former lands.[165]

It is of importance that according to the situational evaluation of the operating NGOs, the government of Bangladesh has recognized the immensity of the challenge of climate displacement for the country. However, it thus far has neither demonstrated the political will nor the financial or technical resources required to implement the rights-based adaptation responses needed to resolve the impending crisis.[166] Consequently, climate displaced communities and persons have received little or no formal support. The absence of targeted government policies and programs may have also opened up the opportunity to work toward generating political will through awareness raising, training and advocacy.

National accommodation and acceptance of the Peninsula Principles in guiding the handling of climate displacement, along with related effective legal and policy solutions, has been reasoned and documented through the achievements of this NGO cooperation. The creation of a Climate Displacement Forum by the Chittagong District Bar Association, for instance, provides the institutional setting to promote awareness of the fundamental rights of climate displaced persons under the Bangladeshi Constitution, to discuss and advocate for legislative change to better protect climate displaced persons and to undertake public interest test cases and strategic litigation on behalf of climate displaced persons guided by the Peninsula Principles.[167]

Finally, ongoing efforts aim at establishing, in partnership with the government, a mechanism for monitoring and recording climate displacement across Bangladesh.

165 See Displacement Solutions, Building on Results: Climate Change and Displacement Initiative (CCDI), Funding Proposal 2014–2016, 2014, p. 49 *et subseq.*, accordingly Displacement Solutions and YPSA have worked with and visited Bangladeshi climate affected communities in Dhaka, Khulna, Chittagong, Rangamati, Bandarban, Cox's Bazaar, Khagrachhari, Jamalpur, Kurigram, Srirajganj, Shadkhira, Jessore, Rajbari, Rangpur, Chandpur and Comilla.

166 *Cf.* Displacement Solutions, Building on Results: Climate Change and Displacement Initiative (CCDI), Funding Proposal 2014–2016, 2014, p. 49.

167 *Ibid.*, p. 52.

Along with this mechanism, a stipulated prioritization of climate displaced people in the redistribution of *khas* land settlements and in case of nonagricultural *khas* land allocation for the settlement of climate displaced peoples should take place in their respective district.[168]

With regard to the rehabilitation of climate displaced persons, there should be a comprehensive rehabilitation policy in place, guided by the Peninsula Principles. Besides the land identification analysis, the idea is for the government and partners to acquire specific land parcels and to bring them into a National Climate Land Bank that will form the institutional basis for ensuring that climate displaced persons and communities are able to obtain new homes and new lands on those *khas*[169] land plots. Land acquisition efforts will be carried out by climate-threatened *upazilas* (local government entities) and will aim to acquire at least 120 ha of land for the relocation of selected climate displaced communities. This will not only have an enormous benefit for the climate displaced persons in Bangladesh, but it could also be an important precedent-setting activity for other countries affected by climate displacement, demonstrating that finding domestic HLP solutions is not only possible, but that it can also provide the basis for a durable solution in handling climate displacement.[170]

## XI IMPLEMENTATION: IMPACT AND IMPEDIMENTS

The Peninsula Principles, eighteen in number, provide guidance to all actors that deal with domestic climate displacement. They provide direction and assist governmental authorities, at the federal, regional or local level, as well as climate displaced persons and communities and their representatives or advocacy organizations, not least international organizations. They apply to all stages of climate displacement. They embody good practices and compile standards for protection against any form of arbitrary displacement, innovatively enunciating a preventive and rights-based land solutions approach with a planned relocation framework for climate displaced persons. The tenets stipulate minimum human rights-based standards and provide

168 Displacement Solutions, The Climate Displacement Land Initiative, Funding Proposal 2016–2020, Climate Justice for the Climate Displaced, 2015, p. 26 *et subseq*.

169 There is currently a large number of "landless" programs implemented by the Government of Bangladesh that are ostensibly designed to provide landless communities with new public land (*khas* land) at no cost. However, despite the clear potential for these programs to benefit climate-displaced communities, there is no specific provision for recognizing or prioritizing climate affected communities among the many categories of landless persons across Bangladesh. Furthermore, these programs are marred by widespread allegations of corruption and mismanagement. Under this action, Displacement Solutions and YPSA will seek to improve the transparency and effectiveness of these landless programmes as well as seek enhanced and guaranteed access of climate displaced persons to receive land under these programmes. See Displacement Solutions, The Climate Displacement Land Initiative, Funding Proposal 2016–2020, Climate Justice for the Climate Displaced, 2015, p. 29.

170 *Ibid.*

guidance for the protection of climate displaced persons within States, applying the full range of human rights from civil and political to economic, social, and cultural to the situation of climate displacement within national borders. Ultimately, the Principles provide for normative benchmarks in conjunction with displacement solutions, for instance through planned relocation processes. A particularity is the *multi-perspectivity* of the Peninsula Principles. It is an attempt to address the full spectrum of stakeholders, whether it is a concerned community or climate displaced person, a local, regional or national authority or a government body with regulatory authority. By taking into account the specific contexts in which these actors operate, and the pressures and constraints they face, the Principles seek to provide a comprehensive set of measures to handle climate displacement based on international minimum standards.

Cognizant of ongoing efforts such as the Nansen Initiative and its particular focus on cross-border climate displacement, the Peninsula Principles concentrated on climate-related internally displaced persons. Domestic climate displacement has two constitutive elements, that is, a person or community needs to be facing or experiencing climate displacement (i.e. be coerced to move while remaining within the national state borders). The Principles address the particular protection gap with regard to internally displaced persons, who do not in all circumstances qualify for protection under existing instruments, standards *et al.* The jurisprudential substance of the Peninsula Principles derives from the concept of sovereignty as a form of responsibility, the compulsory nature of international human rights law and the individual right to HLP law, which has been recognized as one of the non-derogable human rights.

Further to postulating the prime responsibility by governments, their accountability for human rights compliance and the safety of climate displaced persons, the Principles consider it an obligation of the international community to provide humanitarian assistance and protection to displaced persons, if the governments concerned are unable to fulfill their responsibilities. In such instances, governments are supposed to request and accept international bids of aid.

Notwithstanding the fact that the Peninsula Principles were not drafted or formally approved by an intergovernmental process, approval and adherence from UN bodies and agencies, regional organizations, NGOs and a growing number of governments have begun to cite them and use them as the basis for policies, laws and programs for the internally climate displaced.

This receptiveness can be explained with a quest for guidance and pragmatic answers. The Peninsula Principles are plainly written and benefit from systematic and jurisprudential as well as policy coherence. Finally, they address a thematic gap providing concrete guidance to the field and those, for example, operational agencies and NGOs who have to deal with climate displacement on the ground.

Further, the fact that the tenets reflect existing international and human rights law already negotiated and accepted by States reassures many local authorities and line ministries in charge of dealing with climate displacement, thus facilitating their application. The group of eminent jurists, text writers, legal scholars and climate change experts who gathered on the Mornington Peninsula, Australia, on August 18, 2013, did not foresee more than restating norms and law. The innovative defy was to compile existing law and address gaps and gray areas drawing from good practices and suggesting solutions for a coherent approach.

For example, some NGOs urged the drafting committee and group of experts to strengthen the obligations of governments to provide compensation or to address negligence of competent government authorities in detecting or addressing climate displacement, but the Principles do not go beyond applicable existing law and standards.

As a matter of fact, the project was embraced by a large spectrum of actors. It has the key attributes of a multi-stakeholder initiative with cross-sectorial and cross-regional support. The recurring, confirmative and appreciative experience in favor of this initiative from a broad array of stakeholders, allowed the consolidation of institutional or systemic trust in the sense of Niklas Luhmann.[171] In subsumption to the instant case, the donor support for this project can at least be understood as an expression of sympathy for the plight of climate displaced persons and a reflection of a certain common ground for action. Finally, these partnerships exhibit voices of concerns and concrete demands from the climate displacement affected States, governments, regions and peoples and indicate an implicit endorsement of this cause, which in turn sends a global – and rather strong – signal of determination to change the realities of climate displaced persons and is a stepping stone for a concrete call for action.

[171] According to Luhmann, who differentiates between systemic or institutional and individual trust and confidence, the difference lies in the verifiability. As in interpersonal relationships trust is to a limited extent revisable, institutional or systemic trust and confidence builds on far more than interactive communicative processes and plausibility; see N. Luhmann, Vertrauen, Ein Mechanismus der Reduktion sozialer Komplexität, 2014.

# Conclusion

Climate change is one of the biggest challenges of our times. Climate change will drive millions out of their homes in the years and decades to come. Yet global action lags behind.

This research elicited the intricacies of a more than two-decade-long strive toward the recognition of the impact of climate change on human mobility. The lack of agreement over terminology to designate those who are – and will be – on the move due to the adverse effects of climate change is indicative of underlying disagreements as to the causal relationship between climate change and human mobility on the one hand and over the forced *versus* voluntary character of the movement on the other hand. Various organizations, such as the UNHCR or the IOM, which are key actors in this area, have developed working definitions for their own purposes, which squarely fit into the specificities of their respective mandates. These are just two examples of the many definitional attempts, some of which focus on climate/environmental migration, while others look at people on the move through the disaster rather than the climate change lens. Climate displacement, which is the terminology used throughout this study, tries to capture the forced nature of the movement while establishing a direct link to the climate change phenomenon in its various manifestations, including sudden- and slow-onset disasters. While critiques argue that this broad scope would be too unspecific in practice, excluding slow-onset disasters may not address the actual challenge of climate displacement.

Whereas the IPCC, which was created to scientifically assess the climate change phenomenon, in its first assessment report back in 1990 concluded that climate change would lead to massive displacement, it then back-tracked and for several years adopted a more cautious and nuanced approach focusing on vulnerabilities rather than impacts, before finally affirming the climate change–displacement nexus in 2014. The hesitation to affirm this linkage, which is mirrored in the climate change negotiation framework, is also the political expression of the fear that affirming the option of adaptation to cope with the impact of climate change, one of which could be for people to move from their homes, would weaken mitigation efforts. The recognition that "[c]limate change is undeniable" and "unstoppable,"[1] to use the words of the UN Secretary-General, furthered the acceptance that adaptation was needed alongside mitigation. The Climate Change Conference in Cancún (COP 16) in 2010 adopted the Cancún Adaptation Framework (CAF) as part of the Cancún Agreements, explicitly recognizing migration, displacement and planned relocation as forms of adaptation to climate change. This was an important catalyst for the work of many organizations, including Displacement Solutions, to focus on climate displacement.

The affirmation that came about with the Paris Agreement adopted in December 2015, after more than twenty years of negotiations on a binding climate treaty, of the need to respect human rights obligations when addressing climate change was equally slow to materialize. The clear-cut separation between the human rights and the climate change frameworks progressively blurred and the seaming divide narrowed considerably as a result of a two-pronged approach that was actively pursued by a group of States, including some of the most vulnerable to climate change. States such as Bangladesh and the Philippines, who were the driving forces behind the Human Rights Council's action on climate change, persistently pushed for the recognition of the human rights impact of climate change, while others in the climate negotiation arena advocated for the inclusion of human rights considerations in the climate change framework. The consistent cross-referencing of the climate change framework in Human Rights Council resolutions and, vice versa, of the human rights framework in climate change negotiation outcomes was achieved despite the existing reticence, based on the pretended risk of importing politicization into the scientific debate through the human rights angle.

The realities of climate displacement on the ground revealed the lack of operational, policy and normative guidance for States on how to handle such situations and the need to do so in respect of the human rights of the displaced and of host communities. The examination of the existing frameworks and instruments indicated that while they would partially be applicable to situations of climate displacement, none offered comprehensive coverage or adequate guidance to States and climate displaced persons and communities. The UN Guiding Principles on Internal Displacement, which were adopted in 1998, provide a normative framework to address protection challenges in the context of internal displacement, but do not specifically focus on climate displacement. The analysis showed that prevention in the context of climate displacement as opposed to other types of displacement required positive actions of States, an issue that was not addressed in the Guiding Principles. Another set of principles, the UN Principles on Housing and Property Restitution for Refugees and Displaced Persons, better known as the Pinheiro

[1] A. Guterres, Climate Action: Mobilizing the World, Statement at the New York University Stern School of Business, May 30, 2017.

Principles, encapsulates and restates the existing normative framework in relation to a specific set of rights of displaced persons (i.e. housing, land and property rights). While the protection of these rights is also of concern to climate displaced persons, this is but one of the challenges that need to be addressed when handling climate displacement.

The Peninsula Principles on Climate Change within States are the result of a grassroots push. The lack of guidance and support that organizations on the ground first detected was taken up by Displacement Solutions, a lead international NGO in this field, which tried to catalyze the victims' concerns and spearheaded an all-encompassing consultation process and the drafting of a set of principles to bridge the existing gaps. It brought together a group of international experts and practitioners who in 2013 endorsed the Peninsula Principles.

The study showed that the Peninsula Principles do not confer upon States new responsibilities or create new obligations. They are based on existing norms and provide guidance on how existing obligations apply in situations of climate displacement. The compilation of the existing applicable provisions that were scattered across different legal regimes constitutes significant progress for the protection of climate displaced persons. The Peninsula Principles thus respond to a critical need that States and the international community have to address in the years to come.

The Peninsula Principles are intended to support and guide all actors in the search for viable rights-based solutions for the climate displacement challenge that is becoming more pressing by the day. The Principles identify the rights and guarantees relevant to the protection of the internally displaced throughout all phases of displacement. They provide protection against arbitrary displacement, offer a basis for protection and assistance during displacement, and set out guarantees for safe return.

Climate displacement for most of its part will be internal and it was the drafters' deliberate choice to focus the Principles on this type of displacement. One of the Principles' concerns was to cover all types of adverse effects of climate change that could constitute potential displacement triggers, in particular slow-onset disasters. As a result of this broad scope however – which could be one of the Principles' shortcomings – they do not provide any guidance on how to determine whether displacement occurs as a result of environmental degradation or climate change.

The analysis of the Principles outlined two distinct key features: the focus on prevention and the human rights-based approached. Prevention of climate displacement includes climate displacement risk management and may ultimately require planned relocation as opposed to spontaneous and uncontrolled movements, which exacerbate peoples' vulnerabilities and expose them to manifold risks. For relocation to be successful and sustainable, it requires proper preparation and planning, starting with land identification, habitability and use assessments, a participative process toward ensuring the consent of those affected along with legislative action. The Peninsula Principles speak the language of rights and promote a human rights approach – as opposed to a charity approach – to climate displacement. The Principles tailor the existing *acquis* on internal displacement as well as applicable human rights norms to the circumstances faced by climate displaced persons. This allows States to reach a clear understanding of what the obligations mean in practice in the context of climate displacement.

While the Peninsula Principles gained international recognition, they have not formally been endorsed and aside from the international NGO Displacement Solutions, which does extraordinary work to ensure the Principles translate into actual policies and laws on the ground in those countries it is engaged with. The fact that the Principles are not a binding instrument can be considered both a weakness and a strength: a strength as the Principles do not constitute the least common denominator that regularly emerges of an intergovernmental process, but as the product of a bottom-up grassroots initiative, seek to address the actual issues and provide solutions to the real needs. It has also allowed to considerably speed-up the drafting process, which would otherwise, in the case of the negotiation of a binding instrument, have taken many more years.[2]

The Peninsula Principles are however no end in themselves. Unless they are applied and prove to have an actual impact, they are a dead letter. It is therefore even more important to see that those countries that already embrace these Principles use them as a tool to develop their domestic policies and actions. This could create a dynamic for other States to follow suit. Material practice in the implementation of the tenets that started developing will be the first step toward international endorsement.

Convention-*fatigue* seems to be contrasted by the initiative-competition exhibited at the international level. The 2011 Ministerial Conference of UNHCR has not brought about a clear lead agency on the issue and thus opened the door for initiatives by a number of organizations to mushroom. While the enthusiasm is commensurate to the gravity of the challenges, leaving an early mark may also provide a competitive advantage for future discussions on distribution of responsibilities and access to funding, which should not be overlooked as one of several motives. Germany, with its Platform on Disaster Displacement, took the Nansen Initiative on disaster-induced cross-border displacement and its protection agenda under its wings and it remains to be seen how this will further evolve.

The Peninsula Principles are less concerned with this institutional infight. They are complementary to some of these initiatives and have inspired others. They draw their legitimacy from their genesis as a product of grassroots efforts. They are not the creation of an intergovernmental lab but are based on real case scenarios and firsthand information. Their strength is their action-orientation providing step-by-step guidance and solutions to real problems.

[2] *Cf.* K. Koser, Gaps in IDP Protection, 2008, p. 17.

The Principles are being referenced and acknowledged by some of the leading actors in the area of climate displacement, such as Mary Robinson of Climate Justice or UNHCR; others suggested that they form the basis of further research, such as the think tank of the Human Rights Council (i.e. the so-called Advisory Committee).[3] They also underpin action in countries such as Bangladesh. With the recent signing, on October 11, 2017, of a Memorandum of Understanding (MoU) between Displacement Solutions and UN-Habitat, the Peninsula Principles become more engrained in UN action. The MoU provides a framework for cooperation to reduce the impact of climate change and increase resilience of human settlements. The focus is on predictive planning, using the Principles to inform resilience action plans.[4]

The present research does not advocate for the efforts to adopt a binding agreement to be abandoned in the light of the Peninsula Principles. If an internationally recognized instrument on climate displacement were to be adopted, this would obviously provide stronger protection to climate displaced persons and would thus be preferable. Climate change and the ensuing displacement are high on the international agenda. But are the prospects for a binding agreement better today than in 2007 when António Guterres, the then High Commissioner for Refugees and now UN Secretary-General, diagnosed that climate change and environmental degradation were among the three mains reasons for people being on the move, but that there is no political will to address this matter?[5] In other words, is the recognition of the climate change–displacement nexus that has since emerged matched with the required political will to translate pledges into concrete action and regulatory and protection measures?

Louise Arbour's call that it is "time for political leaders to rise to the challenge and not wait to be carried by a wave of supportive public opinion,"[6] shows how well aware the UN Special Representative on International Migration is of the challenges to spur action in a climate of growing nativism in the United States and some European countries, as the agenda remains dominated by the crisis in the Mediterranean and the Balkan routes for Syrian refugees. This dangerous erosion of solidarity and empathy and the growing nationalization of foreign policy directed toward domestic constituencies does not bode well with the rights of people for protection notably under the 1951 Geneva Refugee Convention. "[T]he despair and the urgency … about the need to instil a sense of humanity and evidence-based reasoning into current debates and policy discussions,"[7] voiced by the UNHCR Assistant High Commissioner for Protection, Volker Türk, in his address to the

3 A/HRC/AC/17/2, annex IV.

4 See http://displacementsolutions.org/ds-signs-mou-with-united-nations/ (accessed May 13, 2019).

5 A. Guterres, Keynote speech at the Third Symposium on Corporate Social Responsibility and Humanitarian Assistance, Tokyo, November 26, 2007.

6 J. Bowen, Changing the Narrative on Migration: Q&A with Louise Arbour, September 18, 2017.

7 V. Türk, Statement to the 68th session of the Executive Committee of the High Commissioner's Programme, October 5, 2017.

Executive Committee on October 5, 2017, attests to this effect. He further noted that "[t]he portrayal of the 'other' must start and end with the plight of the human being before us. The orientation towards that plight is the beginning of good governance. The question for all of us is what more can we do so that reason and empathy prevail."[8]

The Global Compact on Safe, Regular and Orderly Migration was adopted in December 2018 and seeks to address all aspects of international migration, including the humanitarian, developmental, human rights-related and other aspects and to develop mechanisms for international cooperation between States of origin and States of destination in this regard. While the Global Compact, together with the Global Compact on Refugees, which was also adopted in December 2018, is forward-looking, it neither comprehensively addresses the issue of internal climate displacement nor constitutes a binding international agreement, which seems to be out of reach in the current climate of rising populism, racism and xenophobia scapegoating migrants. But even if the political landscape were to be more moderate, it has to be borne in mind that no Western country has to date ratified the International Convention on the Protection of Migrant Workers and their Families[9] and that even if an international agreement could be reached against all odds[10] – as the Small Island Developing States (SIDSs) and the Least Developed Countries (LDCs) together hold more than one third of the General Assembly votes – the prospects for its effective implementation would remain grim.[11]

It has also been "pretty explicit that member states did not want internal displacement to feature in this particular compact on migration,"[12] even though the Special Rapporteur on the human rights of internally displaced persons had called for prevention, preparedness and longer-term recovery to be part of it.[13] States continue to remain reluctant to address at the global level or even multilaterally internal displacement, which would make up most of the climate displacement, as it remains a highly politically sensitive matter for reasons of potential interference into domestic affairs.[14] There are discussions on how to strengthen the international protection

8 *Ibid.*

9 For the status of ratifications, see http://indicators.ohchr.org (accessed May 12, 2019).

10 Bearing in mind that the last major UN human rights convention (i.e. the UN Convention on the Rights of Persons with Disabilities (CRPD)) was adopted in 2006 and that this was driven by very particular dynamics influenced by the subject matter and as a result of a sophisticated network of civil society organizations (CSOs) working in concert with a number of States. The example of older persons illustrates that even though there is support of the G77, the African Group and the Group of States of Latin America and the Caribbean (GRULAC) the reticence to enter into new agreements, particularly of States of the Western European and Others Group (WEOG) led to a stalemate on the question of whether or not a new instrument should be elaborated. See A/HRC/33/44, para. 124 *et seq.*

11 See D. Fidler, Disaster Relief and Governance after the Indian Ocean Tsunami: What Role for International Law?, 2005, p. 458.

12 J. Bowen, Changing the Narrative on Migration: Q&A with Louise Arbour, September 18, 2017.

13 A/71/279, para. 12 *et subseq.*

14 A/AC.96/1104, September 20, 2011, para. 33; EC/62/SC/CRP.19, June 6, 2011, para. 4; W. Kälin, From the Nansen Principles to the Nansen Initiative, 2012, p. 49.

framework for internally displaced persons. In this context, the former Special Rapporteur on internally displaced persons, Chaloka Beyani, for instance, recommended to appoint a Special Representative of the Secretary-General on internally displaced persons, with appropriate staff and resources, working within and outside the United Nations and in dialogue with Member States and all stakeholders, is currently being discussed at the international level.[15] "This would strengthen the capacity of the United Nations by establishing a standing office and demonstrate its commitment and that of the international community to addressing internal displacement effectively."[16] It would also complement efforts under the Global Compact with the internal dimension of people on the move and allow to advocate for adequate protection of climate displaced persons within States. In the same vein, several Member States proposed in July 2018 that the Secretary-General appoint a High Level Panel on Internally Displaced Persons, with a diverse and balanced composition and leadership, which could offer an opportunity to galvanize global attention and action on IDPs, noting that past UN-led High Level Panels accelerated progress on a number of comparable challenges where significant obstacles to change or limitations existed within the system in place.

As the Secretary-General noted at the closing of the Katowice Climate Conference:

> From now on, my 5 priorities will be: ambition, ambition, ambition, ambition and ambition.
>
> Ambition in mitigation. Ambition in adaptation. Ambition in finance. Ambition in technical cooperation and capacity building. Ambition in technological innovation.
>
> Ambition will be at the centre of the Climate Summit that I will convene in September 2019.[17]

Meanwhile, it is hoped that human rights mechanisms, national and international NGOs, as well as governments, will further embrace the Peninsula Principles and use them as the basis and source for action in support of those faced with the reality of climate change.

[15] In June 2016 (i.e. the pre-Trump era), the United States expressed support for this recommendation: "We fully agree with his assessment that the UN should restore the position of the Special Representative of the Secretary-General, or SRSG, on the Protection and Assistance of IDPs. An SRSG would have more resources and a more comprehensive mandate than the current position of a Special Rapporteur. This enhancement is necessary to more effectively address the problems confronting IDPs around the world." United States of America, Statement under Agenda Item 3, Clustered Interactive Dialogue with Special Rapporteur on internally displaced persons and the Special Rapporteur on extrajudicial, summary or arbitrary executions, 32nd session of the UN Human Rights Council, June 20, 2016.

[16] A/71/279, para. 93(e).

[17] A. Guterres, Remarks at the Conclusion of COP 24, Katowice, December 15, 2018.

# Bibliography

## ARTICLES


AGA KHAN, Sadruddin: "Legal Problems Relating to Refugees and Displaced Persons," in: *Recueil des Cours de l'Académie du Droit International* 149 (1976), pp. 287–352.

ALFREDSSON, Gudmundur: "Good Governance in the Arctic," in: LOUKACHEVA, Natalia (Ed.), *Polar Law Textbook II*, Copenhagen: Nordic Council of Ministries, 2013, pp. 185–199.

ARENDT, Hannah: "We Refugees," first published in 1943 in *Menorah Journal*, now available in: KOHN, Jerome/ FELDMAN, Ron H. (Eds.), *The Jewish Writings*, New York: Schocken Books, 2007, pp. 264–274.

BARDSLEY, Douglas K./ HUGO, Graeme J.: "Migration and Climate Change: Examining Thresholds of Change to Guide Effective Adaptation Decision-Making," in: *Population and Environment* 32 (2010), pp. 238–262.

BARNETT, Jon/ WEBBER, Michael: "Migration as Adaptation: Opportunities and Limits," in: McADAM, Jane (Ed.), *Climate Change and Displacement: Multidisciplinary Perspectives*, Oxford: Hart, 2012, pp. 37–56.

BATES, Diana C.: "Environmental Refugees? Classifying Human Migrations Caused by Environmental Change," *Population and Environment* 23 (2002), pp. 465–477.

BIERMANN, Frank/ BOAS, Ingrid: "Preparing for a Warmer World: Towards a Global Governance System to Protect Climate Refugees," in: Global Governance Working Paper No. 33 (2007), subsequently published in: *Global Environmental Politics* 10 (2010), pp. 60–88.

BLACK, Richard/ BENNETT, Stephen R. G./ THOMAS, Sandy M./ BEDDINGTON, John R.: "Climate Change: Migration as Adaptation," in: *Nature* 478 (2011), pp. 447–449.

BOEHMER-CHRISTIANSEN, Sonja: "The Precautionary Principle in Germany: Enabling Government," in: O'RIORDAN, Timothy/ CAMERON, James (Eds.), *Interpreting the Precautionary Principle*, London: Earthscan, 1994, pp. 31–60.

BOWEN, James: "Changing the Narrative on Migration: Q&A with Louise Arbour," September 18, 2017. https://theglobalobservatory.org/2017/09/international-migration-refugees-global-compacts-louise-arbour/ (accessed May 10, 2019)

BOYLE, Alan: "Human Rights and the Environment: Where Next?," in: *European Journal of International Law* 23 (2012), pp. 613–642.

BROMWICH, Brendan: "Environmental Degradation and Conflict in Darfur: Implications for Peace and Recovery," in: *Humanitarian Exchange* 39 (2008), pp. 22–29.

BRONEN, Robin: "Alaskan Communities' Rights and Resilience," in: *Forced Migration Review* 31 (2008), pp. 30–32.

"Choice and Necessity: Relocations in the Arctic and South Pacific," in: *Forced Migration Review* 45 (2014), pp. 17–21.

"Climate Displacement: Preparation and Planning," in: LECKIE, Scott/ HUGGINS, Chris (Eds.), *Repairing Domestic Climate Displacement*, London: Routledge, 2016, pp. 89–110.

"Climate-Induced Community Relocations: Creating an Adaptive Governance Framework Based in Human Rights Doctrine," in: *NYU Review of Law and Social Change* 35(2) (2011), pp. 101–148.

"Forced Migration of Alaskan Indigenous Communities Due to Climate Change," in: AFIFI, Tamer/ JÄGER, Jill (Eds.), *Environment, Forced Migration and Social Vulnerability*, London: Springer, 2011, pp. 87–98.

BROWN, Oli: "The Numbers Game," in: *Forced Migration Review* 31 (2008), pp. 8–9.

CAMERON, Edward: "Human Rights and Climate Change: Moving from an Intrinsic to an Instrumental Approach," in: *Georgia Journal of International and Comparative Law* 38 (2010), pp. 673–716.

CEPEDA-ESPINOSA, Manuel José: "How Far May Colombia's Constitutional Court Go to Protect IDP Rights?" in: *Forced Migration Review*, special edition (December 2006), pp. 21–23.

CHAPMAN, Megan S.: "Climate Change and the Regional Human Rights Systems," in: *Sustainable Development Law & Policy* 10 (2009–2010), pp. 37–39.

COHEN, Roberta: "Lessons from the Development of the Guiding Principles on Internal Displacement," in: *Forced Migration Review* 45 (2014), pp. 12–14.

DENMARK, Maria: "Land rights key to peace process," in: *Myanmar Times*, June 10, 2013. www.mmtimes.com/index.php/national-news/7071-land-rights-must-take-central-position-in-peace-process-report-2.html (accessed March 14, 2019)

DOCHERTY, Bonnie: "General Obligations," in: LECKIE, Scott/ HUGGINS, Chris (Eds.), *Repairing Domestic Climate Displacement: The Peninsula Principles*, London: Routledge, 2016, pp. 66–88.

DOCHERTY, Bonnie/ GIANNINI, Tyler: "Confronting a Rising Tide: A Proposal for a Convention on Climate Change Refugees," in: *Harvard Environmental Law Review* 33 (2009), pp. 349–403.

DUN, Olivia/ GEMENNE, Francois: "Defining 'Environmental Migration'," in: *Forced Migration Review* 31 (2008), pp. 10–11.

DUYCKS, Sebastien: "UN Human Rights Council declares climate a priority," *Climate Home News*, June 13, 2016. www.climatechangenews.com/2016/07/13/un-human-rights-council-declares-climate-a-priority/ (accessed May 4, 2019)

EVANS, Judith: "Climate change drives Maldives to buy land," *Financial Times*, November 10, 2008.

EZRA, Markos/ KIROS, Gebre-Egziabher: "Rural Out-Migration in the Drought Prone Areas of Ethiopia: A Multilevel Analysis," in: *International Migration Review* 35 (2011), pp. 749–771.

FEW, Roger/ BROWN, Katrina/ TOMPKINS, Emma L.: "Public Participation and Climate Change Adaptation: Avoiding the Illusion of Inclusion," in: *Climate Policy* 7 (2007), pp. 46–59.

FIDLER, David: "Disaster Relief and Governance after the Indian Ocean Tsunami: What Role for International Law?" in: *Melbourne Journal of International Law* 6 (2005), pp. 458–474.

FISHER, David: "Domestic Regulation of International Humanitarian Relief in Disasters and Armed Conflict: A Comparative Analysis," in: *International Review of the Red Cross* 89 (2007), pp. 345–371.

FISCHER, Frank: "Participatory Governance: From Theory to Practice," in: LEVI-FAUR, David (Ed.), *The Oxford Handbook of Governance*, Oxford: Oxford University Press, 2012, pp. 457–471. www.oxfordhandbooks.com/view/10.1093/oxfordhb/9780199560530.001.0001/oxfordhb-9780199560530-e-32 (accessed May 16, 2019)

FUSSELL, Elizabeth: "Space, Time and Volition: Dimensions of Migration Theory," in: ROSENBLUM, Marc R./ TICHENOR, Daniel J. (Eds.), *The Oxford Handbook of the Politics of International Migration*, Oxford: Oxford University Press, 2012, pp. 25–52.

GIBB, Christine/ FORD, James: "Should the United Nations Framework Convention on Climate Change Recognize Climate Migrants?," in: *Environmental Research Letters* 7 (2012), pp. 1–9.

GRAY, Clark/ BILSBORROW, Richard: "Environmental Influences on Human Migration in Rural Ecuador," in: *Demography* 50(4) (2013), pp. 1–25.

GULLETT, Warwick: "The Threshold Test of the Precautionary Principle in Australian Courts and Tribunals: Lessons for Judicial Review," in: FISHER, Elizabeth C./ JONES, Judith S./ VON SCHOMBERG, René (Eds.), *Implementing the Precautionary Principle: Perspectives and Prospects*, Cheltenham: Edward Elgar, 2005, pp. 182–201.

HAIGH, Nigel: "The Introduction of the Precautionary Principle in the UK," in: O'RIORDAN, Timothy/ CAMERON, James (Eds.), *Interpreting the Precautionary Principle*, London: Earthscan, 1994, pp. 229–251.

HALL, Nina: "Climate Change and Organizational Change in UNHCR," in: LEIGHTON, Michelle/ SHEN, Xiaomeng/ WARNER, Koko (Eds.), *Climate Change and Migration: Rethinking Policies for Adaptation and Disaster Risk Reduction*, Bonn: United Nations University, 2011, pp. 102–112.

HODGKINSON, David: "A Brief Overview of the Drafting of the Peninsula Principles," in: LECKIE, Scott/ HUGGINS, Chris (Eds.), *Repairing Domestic Climate Displacement: The Peninsula Principles*, London: Routledge, 2016, pp. 29–35.

HODGKINSON, David/ BURTON, Tess/ ANDERSON, Heather/ YOUNG, Lucy: "'The Hour When the Ship Comes In': A Convention for Persons Displaced by Climate Change," in: *Monash University Law Review* 36(1) (2010), pp. 1–49.

HÖLL, Otmar: "Die Ergebnisse der UN-Umweltkonferenz von Rio (3.–14. Juni 1992)," in: CALLIESS, Jörg (Ed.), *Die transatlantischen Beziehungen im Umbruch, Loccumer Protokolle* 55/92, 1992, pp. 261–268.

"Ökologisierung der internationalen Politik als Konfliktpotential: Die Konferenz über Umwelt und Entwicklung von Rio und die Folgen," in: *Friedensbericht* (Dialog 1–2/1993), pp. 187–201.

HOMER-DIXON, Thomas F.: "On the Threshold: Environmental Changes as Causes of Acute Conflict," in: *International Security* 16(2) (1991), pp. 76–116.

HUGO, Graeme: "Environmental Concerns and International Migration," in: *International Migration Review* 30 (1996), pp. 105–131.

KÄLIN, Walter: "From the Nansen Principles to the Nansen Initiative," in: *Forced Migration Review* 41 (2012), pp. 48–49.

"Turning the Vision of an Effective Human Rights Council into Reality," in: MÜLLER, Lars (Ed.), *The First 365 Days of the United Nations Human Rights Council*, Geneva, 2006, pp. 225–233.

KIBREAB, Gaim: "Environmental Causes and Impact of Refugee Movements: A Critique of the Current Debate," in: *Disasters* 21(1) (1997), pp. 20–38.
KLEIN SOLOMON, Michele/ WARNER, Koko: "Protection of Persons Displaced as a Result of Climate Change: Existing Tools and Emerging Frameworks," in: GERRARD, Michael B./ WANNIER, Gregory E. (Eds.), *Threatened Island Nations: Legal Implications of Rising Seas and a Changing Climate*, Cambridge: Cambridge University Press, 2013, pp. 243–298.
KNOX, John: "Climate Change and Human Rights Law," in: *Virginia Journal of International Law* 50(1) (2009), pp. 164–218.
KOLMANNSKOG, Vikram: "Climate Change, Human Mobility, and Protection: Initial Evidence from Africa," in: *Refugee Survey Quarterly* 29(3) (2010), pp. 103–119.
KOSER, Khalid: "Gaps in IDP Protection," in: *Forced Migration Review* 31 (2008), p. 17.
KUNZ, Egon: "The Refugees in Flight: Kinetic Models and Forms of Displacement," in: *International Migration Review* 7 (1973), pp. 125–146.
LACHMAYER, Konrad: "The Austrian Constitutional Court," in: JAKAB, Andràs/ DYEVRE, Arthur/ ITZCOVICH, Giulo (Eds.), *Comparative Constitutional Reasoning*, Cambridge: Cambridge University Press, 2017, pp. 36–74.
LECKIE, Scott: "An Integrated Approach to Climate Change Demands That Human Rights and Adaptation Strategies Are Pursued Hand-in-Hand," in: *Forced Migration Review* 31 (2008), pp. 18–19.
"Human Rights Implications," in: *Forced Migration Review* 31 (2008), pp. 18–19.
"Leader of the Pack: Who Will Take the Lead on Post-Conflict HLP Issues," in: PANTULIANO, Sara (Ed.), *Unchartered Territory, Land, Conflict and Humanitarian Action*, Rugby, UK: Practical Action, 2009, pp. 95–106.
"The Great Land Theft," in: *Forced Migration Review*, Special Issue: Tsunami – Learning from the Humanitarian Response (2005), pp. 15–16.
"Using Human Rights to Resolve the Climate Displacement Problem: The Promise of the Peninsula Principles," in: LECKIE, Scott/ HUGGINS, Chris (Eds.), *Repairing Domestic Climate Displacement: The Peninsula Principles*, London: Routledge, 2016, pp. 1–15.
LEWIS, Dan A.: "The Relocation of Development from Coastal Hazards through Publicly Funded Acquisition Programs: Examples and Lessons from the Gulf Coast," in: *Sea Grant Law and Policy Journal* 5(1) (2012), pp. 98–139.
LIMON, Marc: "Human Rights and Climate Change: Constructing a Case for Political Action," in: *Harvard Environmental Law Review* 33 (2009), pp. 439–476.
LONERGAN, Steve: "The Role of Environmental Degradation in Population Displacement," in: *Environmental Change and Security Project Report* 4 (1998), pp. 5–15.
LOUGHRY, Maryanne/ McADAM, Jane: "Kiribati: Relocation and Adaptation," in: *Forced Migration Review* 31 (2008), pp. 51–52.
MARCHILDON, Gregory P./ KULSHRESHTA, Suren/ WHEATON, Elaine/ SAUCHYN, Dave: "Drought and Institutional Adaptation in the Great Plains of Alberta and Saskatchewan, 1914–1939," in: *Natural Hazards* 45 (2008), pp. 391–411.
MARTIN, Susan F.: "Climate Change, Migration and Governance," in: *Global Governance* 16 (2010), pp. 397–414.
"Managing Environmentally Induced Migration," in: LACKO, Frank/ AGHAZARM, Christine (Eds.), *Migration, Environment and Climate Change: Assessing the Evidence*, Geneva: IOM, 2009, pp. 355–384.

MAY, Bradley/ PLUMMER, Ryan: "Accommodating the Challenges of Climate Change Adaptation and Governance in Conventional Risk Management: Adaptive Collaborative Risk Management (ACRM)," in: *Ecology and Society* 16 (2011), pp. 47–62.

McADAM, Jane: "Creating New Norms on Climate Change, Natural Disasters and Displacement: International Developments 2010–2013," *Refuge* 29(2) (2014), pp. 11–26.

"'Disappearing States,' Statelessness and the Boundaries of International Law," in: McADAM, Jane (Ed.), *Climate Change and Displacement, Multidisciplinary Perspectives*, Oxford: Hart, 2012, pp. 105–129.

"Swimming against the Tide: Why a Climate Change Displacement Treaty Is Not the Answer," in: *International Journal for Refugee Law* 23 (2011), pp. 2–27.

McANANEY, Sheila C.: "Sinking Islands? Formulating a Realistic Solution to Climate Change Displacement," in: *New York University Law Review* 87 (2012), pp. 1172–1209.

McINERNEY-LANKFORD, Siobhàn: "Human Rights and Climate Change: Reflections on International Legal Issues and Potential Policy Relevance," in: GERRARD, Michael B./ WANNIER, Gregory E. (Eds.), *Threatened Island Nations: Legal Implications of Rising Seas and a Changing Climate*, Cambridge: Cambridge University Press, 2013, pp. 195–242.

MYERS, Norman: "Environmental Refugees: A Growing Phenomenon of the 21st Century," in: *Philosophical Transactions: Biological Sciences* 357 (2002), pp. 609–613.

"Environmental Refugees in a Globally Warmed World," in: *Bioscience*, 43 (1993), pp. 752–761.

O'RIORDAN, Timothy/ JORDAN, Andrew: "The Precautionary Principle in Contemporary Environmental Politics," in: *Environmental Values* 4 (1995), pp. 191–212.

PENDER, James: "Community-Led Adaptation in Bangladesh," in: *Forced Migration Review* 31 (2008), pp. 54–55.

PERCH-NIELSEN, Sabine/ BÄTTIG, Michèle B./ IMBODEN, Dieter: "Exploring the Link between Climate Change and Migration," in: *Climatic Change* 91 (2008), pp. 375–393.

PERRY, Ronald W.: "What Is a Disaster?," in: RODRÍGUEZ, Havidán/ QUARANTELLI, Enrico L./ DYNES, Russell R. (Eds.), *Handbook of Disaster Research*, New York: Springer, 2007, pp. 1–15.

PIELKE, Roger A. Jr.: "Misdefining 'Climate Change': Consequences for Science and Action," in: *Environmental Science & Policy* 8 (2005), pp. 548–561.

PIGUET, Etienne/ PÉCOUD, Antoine/ DE GUCHTENEIRE, Paul: "Migration and Climate Change: An Overview," in: *Refugee Survey Quarterly* 30(3) (2011), pp. 1–23.

RICHMOND, Anthony H.: "Reactive Migration: Sociological Perspectives on Refugee Movements," in: *Journal of Refugee Studies* 6 (1993), pp. 7–24.

"Sociological Theories of International Migration: The Case of Refugees," in: *Current Sociology* 36(2) (1988), pp. 7–25.

RIVERA, Humberto Cantú: "The United Nations Human Rights Council: Remarks on Its History, Procedures, Challenges and Perspectives," in: RIVERA, Humberto Cantú (Ed.), *The Special Procedures of the Human Rights Council: A Brief Look from the Inside and Perspectives from the Outside*, Cambridge: Intersentia, 2015, pp. 839–843.

ROBINSON, Mary: "What Rights Can Add to Good Development Practice," in: ALSTON, Philip/ ROBINSON, Mary (Eds.), *Human Rights and Development: Towards Mutual Reinforcement*, Oxford: Oxford University Press, 2005. www.oxfordscholarship.com/view/10.1093/acprof:oso/9780199284627.001.0001/acprof-9780199284627 (accessed April 20, 2019)

ROTHRING, Jacob/ ROMERO, Marco: "Measuring the Enjoyment of Rights in Colombia," in: *Forced Migration Review* 30 (2008), pp. 64–65.

SACHS, Wolfgang: "Climate Change and Human Rights," in: *Development* 51 (2008), pp. 332–337.

SCHIPPER, Lisa: "Conceptual History of Adaptation in the UNFCCC Process," in: *Review of European, Comparative and International Environmental Law* 15 (2006), pp. 82–92.

SCHÜTZ-MÜLLER, Ingfrid: "Excursions to UN Headquarters and Specialized Agencies: Report on Experiences," in: *Can the United Nations Be Taught?* Vienna: Diplomatic Academy, 2009, pp.106–107.

"Von der Entkolonisierung zu den gegenwärtigen Problemen der pazifischen Inselstaaten," in: MÜCKLER, Hermann/ ORTMAYR, Norbert/ WERBER, Harald (Ed.), *Ozeanien: 18. bis 20. Jahrhundert: Geschichte und Gesellschaft*, Vienna: Promedia, 2009, pp. 286–302.

"Zur Funktion von Intellektuellen in pazifischen Inselstaaten," in: KREISKY, Eva (Ed.), *Von der Macht der Köpfe, Intellektuelle zwischen Moderne und Spätmoderne*, Vienna: WUV, 2000, pp. 312–321.

SIMEL, Joseph Ole: "Le pastoralisme et les defis du changement climatique," in: *Questions Autochtones* 3–4/09 (2009), pp. 30–37.

SIMPERINGHAM, Ezekiel: "The Responsibilities of States to Protect Climate-Displaced Persons," in: LECKIE, Scott/ HUGGINS, Chris (Eds.), *Repairing Domestic Climate Displacement: The Peninsula Principles*, London: Routledge, 2016, pp. 111–137.

SINDICO, Francesco: "Climate Change: A Security (Council) Issue?" in: *Climate Change Law Review* 1 (2007), pp. 29–34.

STEPHENS, Pamela J.: "Applying Human Rights Norms to Climate Change: The Elusive Remedy," in: *Colorado Journal of International Environmental Law and Policy* 21 (2010), pp. 49–83.

SUHRKE, Astri: "Environmental Degradation and Population Flows," in: *Journal of International Affairs* 47 (1994), pp. 473–496.

TACOLI, Cecilia: "Crisis or Adaptation? Migration and Climate Change in a Context of High Mobility," in: *Environment and Urbanization* 21 (2009), pp. 513–525.

ULLMAN, Richard H.: "Redefining Security," in: *International Security* 8 (1983), pp. 129–153.

VON SCHOMBERG, René: "The Precautionary Principle and Its Normative Challenges," in: FISHER, Elizabeth Charlotte/ JONES, Judith S./ VON SCHOMBERG, René (Eds.), *Implementing the Precautionary Principle: Perspectives and Prospects*, Cheltenham: Edward Elgar, 2005, pp. 19–41.

WARNER, Koko: "Global Environmental Change and Migration: Governance Challenges," in: *Global Environmental Change* 20 (2010), pp. 402–413.

WARNER, Koko/ LACZKO, Frank: "A Global Research Agenda," in: *Forced Migration Review* 31 (2008), pp. 59–61.

WILLIAMS, Angela: "Turning the Tide: Recognizing Climate Change Refugees in International Law," *Law & Policy* 30 (2008), pp. 502–529.

WILLIAMS, Rhodri C.: "Property," in: KÄLIN, Walter/ WILLIAMS, Rhodri C/ KOSER, Khalid/ SOLOMON, Andrew (Eds.), *Incorporating the Guiding Principles on Internal Displacement into Domestic Law: Issues and Challenges*, Studies in Transnational Legal Policy, No. 41, Washington DC: American Society of International Law, 2010, pp. 363–432.

WYMAN, Katrina Miriam: "Responses to Climate Migration," in: *Harvard Environmental Law Review* 37 (2013), pp. 167–216.

ZETTER, Roger: "Legal and Normative Frameworks," in: *Forced Migration Review* 37 (2008), pp. 62–63.

"The Role of Legal and Normative Frameworks for the Protection of Environmentally Displaced Persons," in: LACKO, Frank/ AGHAZARM, Christine (Eds.): *Migration, Environment and Climate Change: Assessing the Evidence*, IOM: Geneva, 2009, pp. 385–441.

## MONOGRAPHS

ARENDT, Hannah: *Wir Flüchtlinge*, Mit einem Essay von Thomas Meyer, 5th ed., Ditzingen: Reclam Universal Bibliothek, 2016.

AUGENEDER, Silvia: "The Denial of Human Dignity from a Legal and Philosophical Point of View," in: PECZENIK, Aleksander (Ed.), *Proceedings of the 21st IVR World Congress, Lund (Sweden), 12–17 August 2003, Part 1: Justice*, pp. 193–200, Archives for Philosophy of Law and Social Philosophy, Stuttgart: Franz Steiner, 2004.

BARNETT, Jon: Security and Climate Change, Tyndall Centre Working Paper 7. Norwich: Tyndall Centre for Climate Change Research, 2001.

BARNETT, Jon/ WEBBER, Michael: Accommodating Migration to Promote Adaptation to Climate Change, World Bank, Research Paper 5270, 2010.

BLACK, Richard: Environmental Refugees: Myth or Reality?, New Issues in Refugee Research, Research Paper No. 34, Geneva: UNHCR, 2001. www.unhcr.org/3ae6a0d00.html (accessed February 18, 2019)

BRONEN, Robin/ CHAPIN, Stuart F. : Adaptive Governance and Institutional Strategies for Climate-Induced Community Relocations in Alaska. Proceedings of the National Academy of Sciences, Washington, DC: National Academy of Sciences, 2013.

BROWN, Lester R.: *World Population Trends: Signs of Hope, Signs of Stress*, Washington, DC: World Watch Institute, 1976.

BROWN, Oli: Migration and Climate Change, IOM Migration Research Series No. 30, Geneva: IOM, 2008.

*Weather of Mass Destruction? The Rise of Climate Change as a "New" Security Issue*, Winnipeg: International Institute for Sustainable Development, 2007.

CASTLES, Stephen: Environmental Change and Forced Migration: Making Sense of the Debate, New Issues in Refugee Research, Research Paper No. 70, Geneva: UNHCR, 2002. www.unhcr.org/3de344fd9.html (accessed February 18, 2019)

CLARK, William A. V.: *Environmentally Induced Migration and Conflict, External Expertise for the WBGU Report World in Transition: Climate Change as a Security Risk*, Berlin: German Advisory Council on Global Change, 2006.

DE SCHUTTER, Olivier: *International Human Rights Law: Cases, Materials, Commentary*, Cambridge: Cambridge University Press, 2014.

DUCASSE-ROGIER, Marianne, *The International Organization for Migration, 1951–2001*, Geneva: International Organization for Migration, 2002.

DuNANN WINTER, John: *An Introduction to Igneous and Metamorphic Petrology*, Upper Saddle River, NJ: Prentice Hall, 2001.

EL-HINNAWI, Essam: *Environmental Refugees*, Nairobi: UNEP, 1985.

ERMACORA, Felix: *Handbuch der Grundfreiheiten und Menschenrechte in Österreich*, Vienna: Manz, 1963.

FERRIS, Elizabeth G.: *The Politics of Protection: The Limits of Humanitarian Action*, Washington DC: Brookings, 2011.

FORSYTH, David. P.: *Human Rights in International Relations*, Cambridge: Cambridge University Press, 2012.
FREEDMAN, Rosa: *The United Nations Human Rights Council: A Critique and Early Assessment*, London: Routledge, 2013.
GLEDITSCH, Nils. P./ NORDÅS, Ragnhild/SALEHYAN, Idean: Climate Change and Conflict: The Migration Link, Coping with Crisis, Working Paper Series, New York: International Peace Academy, 2007.
GLUCK, Jason: Why Sudan's Popular Consultation Matters, United States Institute of Peace, Special Report 260, November 2010.
GÜNDOĞDU, Ayten: *Rightlessness in Age of Rights, Hannah Arendt and the Contemporary Struggles of Migrants*, New York: Oxford University Press, 2015.
HALL, Nina: *Displacement, Development, and Climate Change*, London: Routledge, 2016.
HASLINGER, Josef: *Tsunami: A Report from Phi Phi Island*, California: Ariadne Press, 2007.
HASSELL, James E.: "Russian Refugees in France and the United States between the World Wars," *Transactions of the American Philosophical Society* 81(7), 1991, pp. 1–96.
HASSINE, Khaled: *Housing and Property Directorate/Claims Commission in Kosovo*, Vienna: NWV, 2009.
*Regularizing Property Rights in Kosovo and Elsewhere*, Köln: WiKu, 2010.
HASSINE, Khaled/ LECKIE, Scott: *The United Nations Principles on Housing and Property Restitution for Refugees and Displaced Persons: A Commentary*, International Studies in Human Rights, Vol. 114, Leiden: Brill Nijhoff, 2016.
HASSOL, Susan Joy: *Arctic Climate Impact Assessment (ACIA)*, Cambridge: Cambridge University Press, 2004.
HUGO, Graeme: Migration, Development and Environment, IOM Migration Research Series No. 35, Geneva: IOM, 2008.
JACOBSON, Jodi L.: Environmental Refugees. A Yardstick of Habitability, Worldwatch Paper 86, Washington, DC: Worldwatch Institute, 1988.
JONSSON, Gunvor: The Environmental Factor in Migration Dynamics: A Review of African Case Studies, Working Paper No. 21, International Migration Institute, University of Oxford, 2010. www.imi.ox.ac.uk/pdfs/imi-working-papers/wp21-jonsson (accessed April 27, 2019)
KÄLIN, Walter: Displacement Caused by the Effects of Climate Change: Who Will Be Affected and What Are the Gaps in the Normative Framework for Their Protection?, Background Paper, 2008.
Guiding Principles on Internal Displacement: Annotations, The Brookings Institution – University of Bern Project on Internal Displacement Studies in Transnational Legal Policy No. 38, Washington, DC: American Society of International Law, revised ed., 2008.
KNIVETON, Dominic/ SCHMIDT-VERKERK, Kerstin/ SMITH, Christopher/ BLACK, Richard: *Climate Change and Migration: Improving Methodologies to Estimate Flows*, Migration Research Series No. 33, Geneva: IOM, 2008.
KRAMER, Helmut/ DZIHIC, Vedran: *Die Kosovo Bilanz: Scheitert die internationale Gemeinschaft*, Vienna: LIT, 2005.
LECKIE, Scott (Ed.): *Housing, Land, and Property Rights in Post-Conflict United Nations and Other Peace Operations: A Comparative Survey and Proposal for Reform*, Cambridge: Cambridge University Press, 2009.
(Ed.): *Land Solutions for Climate Displacement*, New York: Routledge, 2014.
(Ed.): *Returning Home: Housing and Property Restitution Rights of Refugees and Displaced Persons*, Ardsley, NY: Transnational Publishers, 2003.

Climate-Related Disasters and Displacement: Homes for Homes, Lands for Lands, Paper presented to the Expert-Group meeting on Population Dynamics and Climate Change, London, June 2009.

*Finding Land Solutions to Climate Displacement: A Challenge Like Few Others*, Geneva: Displacement Solutions, 2013.

LECKIE, Scott/ HUGGINS, Chris (Eds.): *Conflict and Housing, Land and Property Rights: A Handbook on Issues, Frameworks, and Solutions*, Cambridge: Cambridge University Press, 2011.

*Repairing Domestic Climate Displacement: The Peninsula Principles*, London: Routledge, 2016.

LIM, Bo/ SPANGER-SIEGFRIED, Erika (Eds.): *Adaptation Policy Frameworks for Climate Change: Developing Strategies, Policies and Measures*, Cambridge: Cambridge University Press, 2004.

LIMON, Marc/ PICCONE, Ted: *Human Rights Special Procedures: Determinants of Influence, Understanding and Strengthening the Effectiveness of the UN's Independent Human Rights Experts*, Versoix: Universal Rights Group, 2014.

LUHMANN, Niklas: *Vertrauen, Ein Mechanismus der Reduktion sozialer Komplexität*, 5th ed., Konstanz: UTB, 2014.

McADAM, Jane/ LIMON, Marc: *Human Rights, Climate Change and Cross-Border Displacement: The Role of the International Human Rights Community in Contributing to Effective and Just Solutions*, Versoix: Universal Rights Group, 2015.

McLEMAN, Robert: *Climate Change, Migration and Critical International Security Considerations*, IOM Migration Research Series No. 42, Geneva: IOM, 2011.

MISEREZ, Diana: *Prince Sadruddin Aga Khan: Humaniste et Visionnaire*, Bière: Cabédita Editions, 2017.

MORRISSEY, James: *Environmental Change and Forced Migration: A State of the Art Review*, Oxford: Oxford Department of International Development, Refugee Studies Center, 2009.

MYERS, Norman: Environmental Refugees: An Emergent Security Issue, 13th Economic Forum, Prague, May 2005.

MYERS, Norman/ KENT, Jennifer: *Environmental Exodus: An Emergent Crisis in the Global Arena*, Washington DC: Climate Institute, 1995.

NAIK, Asmita/ STIGTER, Elca/ LACZKO, Frank: *Migration, Development and Natural Disasters: Insights from the Indian Ocean Tsunami*, IOM Migration Research Series No. 31, Geneva: IOM, 2007.

NOBBE, Clara: Universality, Common but Differentiated Responsibilities and the Sustainable Development Goals, Working Paper, Division Global Issues, Berlin: Stiftung Wissenschaft und Politik, 2014.

NOWAK, Manfred: *UN Covenant on Civil and Political Rights: CCPR Commentary*, 2nd revised ed., Kehl: N. P. Engel, 2005.

PARK, Susin: *Climate Change and the Risk of Statelessness: The Situation of Low-Lying Island States*, Legal and Protection Policy Research Series, Geneva: UNHCR, 2011.

PAUW, Pieter/ BAUER, Steffen/ RICHERZHAGEN, Carmen/ BRANDI, Clara/ SCHMOLE, Hanna: Different Perspectives on Differentiated Responsibilities: A State-of-the-Art Review of the Notion of Common but Differentiated Responsibilities in International Negotiations, Discussion Paper 6, Bonn: Deutsches Institut für Entwicklungspolitik (DIE), 2014.

PIGUET, Etienne: Climate Change and Forced Migration: How Can International Policy Respond to Climate Change Induced Displacement, New Issues in Refugee Research, Research Paper No. 153, Geneva: UNHCR, 2008.

RALEIGH, Clionadh/ JORDAN, Lisa/ SALEHYAN, Idean: Assessing the Impact of Climate Change on Migration and Conflict, World Bank Seminar on Exploring the Social Dimensions of Climate Change, Washington DC: The World Bank, 2008. http://sitere sources.worldbank.org/EXTSOCIALDEVELOPMENT/Resources/SDCC WorkingPaper_MigrationandConflict.pdf (accessed March 20, 2019)

RAMCHARAN, Bertrand G.: *Preventive Human Rights Strategies*, London: Routledge, 2010.

*The Protection Role of UN Human Rights Special Procedures*, Leiden: Nijhoff, 2009.

*The United Nations High Commissioner for Human Rights: The Challenges of International Protection*, The Hague: Nijhoff, 2002.

RENAUD, Fabrice/ BOGARDI, Janos J./ DUN, Olivia/ WARNER, Koko: *Control, Adapt or Flee: How to Face Environmental Migration?*, Intersections 5, Bonn: UNU Institute for Environment and Human Security (UNU-EHS), 2007.

RODIN, Judith: *The Resilience Dividend: Managing Disruption, Avoiding Disaster, and Growing Stronger in an Unpredictable World*, New York: Profile Books, 2015.

SCHAMBECK, Herbert: *Der Begriff der „Natur der Sache": Ein Beitrag zur rechtsphilosophischen Grundlagenforschung*, Vienna: Springer, 1964.

SIMMA, Bruno/ KHAN, Daniel-Erasmus/ NOLTE, Georg/ PAULUS, Andreas (Eds.), *The Charter of the United Nations: A Commentary*, 3rd ed., Vols. 1 and 2, Oxford: Oxford University Press, 2012. (Mode of citation: contributor, Art., Chapter, para., in: B. Simma *et al.*, The Charter of the United Nations, Vol. 1/2, 2012.)

STERN, Nicolas (Ed.): *The Economics of Climate Change: The Stern Review*, London: HM Treasury, 2006.

WARNER, Koko: Climate Change Induced Displacement: Adaptation Policy in the Context of the UNFCCC Climate Negotiations, May 2011. www.iom.int/cms/climateandmigra tion# (accessed May 27, 2019)

WATT-CLOUTIER, Sheila: *The Right to Be Cold: One Woman's Story of Protecting Her Culture, the Arctic and the Whole Planet*, Canada: Penguin Books, 2016.

WELZER, Harald: *Klimakrieg, Wofür im 21. Jahrhundert getötet wird*, Frankfurt am Main: Fischer Taschenbuch, 2010.

WENTZ, Jessica/ BURGER, Michael: Designing a Climate Change Displacement Coordination Facility: Key Issues for COP 21, Columbia Law School: Sabin Center for Climate Change Law, 2015.

ZETTER, Roger: *Protecting Forced Migrants, A State of the Art Report of Concepts, Challenges and Ways Forward*, Bern: Federal Commission on Migration, 2014.

## SOURCES/PRIMARY DOCUMENTS

A/59/565: Report of the High-Level Panel on Threats, Challenges and Change, A more secure world: Our shared responsibility, December 2, 2004.

A/64/255: Report of the Special Rapporteur on adequate housing as a component of the right to an adequate standard of living and on the right to non-discrimination in this context to the UN General Assembly, Raquel Rolnik, August 6, 2009.

A/66/270: Report of the Special Rapporteur on adequate housing as a component of the right to an adequate standard of living and on the right to non-discrimination in this context to the UN General Assembly, Raquel Rolnik, August 5, 2011.

A/66/285: Report of the Special Rapporteur on the human rights of internally displaced persons to the General Assembly, Chaloka Beyani, Protection of and assistance to internally displaced persons, August 9, 2011.

A/67/299: Report of the Special Rapporteur on the human rights of migrants, François Crépeau, to the UN General Assembly, August 13, 2012.
A/70/287: Interim report of the Special Rapporteur on the right to food, Hilal Elver, to the UN General Assembly, August 5, 2015.
A/71/279: Report of the Special Rapporteur on the human rights of internally displaced persons, August 3, 2016.
A/AC.96/860: Report of the 46th session of the Executive Committee of the High Commissioner's Programme (Geneva, October 16–20, 1995), October 23, 1995.
A/AC.96/199: United Nations High Commissioner for Refugees (UNHCR): Note on international protection submitted to the Executive Committee of the High Commissioner's Programme, 1992.
A/AC.96/1104: Report of the 51st meeting of the Standing Committee, Executive Committee of the High Commissioner's Programme (Geneva, June 21–23, 2011).
A/CN.4/Rev.1: Survey of international law in relation to the work of the International Law Commission, 1949.
A/CN.4/L.871: Titles and texts of the preamble and draft articles 1 to 18 of the draft articles on the Protection of persons in the event of disaster, International Law Commission, 68th session, Geneva, May 2–June 10 and July 4–August 12, 2016.
A/CONF.151/26 (Vol. 1): Report of the United Nations Conference on Environment and Development (Rio de Janeiro, June 3–14, 1992).
A/CONF.157/23: Vienna Declaration and Programme of Action as adopted by the World Conference on Human Rights on June 25, 1993, World Conference on Human Rights (Vienna, June 14–25, 1993), July 12, 1993.
A/CONF.224/CRP.1: Sendai Framework for Disaster Risk Reduction 2015–2030, March 18, 2015.
A/CONF.231/3: Intergovernmental Conference to Adopt the Global Compact for Safe, Orderly and Regular Migration (Marrakech, Morocco, December 10–11, 2018), Item 10 of the Provisional Agenda, Draft Outcome Document of the Conference, Note by the President of the General Assembly, July 30, 2018.
A/HRC/10/7/Add.4: Report of the Special Rapporteur on adequate housing as a component of the right to an adequate standard of living, and on the right to non-discrimination in this context, Raquel Rolnik; Addendum: Summary of Communications Sent and Replies Received from Governments and Other Actors, February 17, 2009.
A/HRC/10/61: Report of the Office of the United Nations High Commissioner for Human Rights on the relationship between climate change and human rights, January 15, 2009.
A/HRC/12/21: Report of the Special Rapporteur on contemporary forms of slavery, including its causes and consequences, Gulnara Shahinian, July 10, 2009.
A/HRC/13/20/Add.3: Report of the Special Rapporteur on adequate housing as a component of the right to an adequate standard of living, and on the right to non-discrimination in this context, Raquel Rolnik; Addendum: Mission to Maldives (February 18–26, 2009), January 11, 2010.
A/HRC/13/21: Report of the Representative of the Secretary-General on the human rights of internally displaced persons, Walter Kälin, January 5, 2010.
A/HRC/16/42: Report of the Special Rapporteur on adequate housing as a component of the right to an adequate standard of living, and on the right to non-discrimination in this context, Raquel Rolnik, December 20, 2010.
A/HRC/16/43: Report of the Special Rapporteur on the human rights of internally displaced persons, Chaloka Beyani, December 20, 2010.

A/HRC/16/62: Human Rights Council Report of the 2010 Social Forum (Geneva, October 4–6, 2010), January 4, 2011.
A/HRC/20/7: Report of the United Nations High Commissioner for Human Rights on the outcome of the seminar addressing the adverse impacts of climate change on the full enjoyment of human rights, April 10, 2012.
A/HRC/22/43: Report of the Independent Expert on the issue of human rights obligations relating to the enjoyment of a safe, clean, healthy and sustainable environment, John H. Knox, December 24, 2012.
A/HRC/30/43/Add.3: Report of the Independent Expert on the enjoyment of all human rights by older persons, Mission to Mauritius, August 12, 2015.
A/HRC/31/39, Annex 11: Report of the High Commissioner for Human Rights on the 22nd Annual Meeting of Special Rapporteurs/Representatives, Independent Experts and Working Groups of the Special Procedures of the Human Rights Council, including updated information on the Special Procedures, February 17, 2016.
A/HRC/31/52: Report of the Special Rapporteur on the issue of human rights obligations relating to the enjoyment of a safe, clean, healthy and sustainable environment, February 1, 2016.
A/HRC/33/44: Report of the Independent Expert on the enjoyment of all human rights by older persons ("Comprehensive Report"), August 4, 2016.
A/HRC/33/44/Add.1: Report of the Independent Expert on the enjoyment of all human rights by older persons on her mission to Costa Rica, July 8, 2016.
A/HRC/4/18, Annex 1: Report of the Special Rapporteur on adequate housing as a component of the right to an adequate standard of living, Basic principles and guidelines on development-based evictions and displacement, February 5, 2007.
A/HRC/AC/17/2: Human Rights Council Advisory Committee, 17th session, August 8–12, 2016, Agenda item 4, Report of the Advisory Committee on its 17th session, Rapporteur: Carla Hananía de Varela, September 7, 2016.
A/HRC/RES/10/4: Human Rights Council resolution of March 25, 2009, "Human rights and climate change."
A/HRC/RES/13/17: Human Rights Council resolution of March 25, 2010, "The Social Forum."
A/HRC/RES/14/6: Human Rights Council resolution of June 17, 2010, "Mandate of the Special Rapporteur on the human rights of internally displaced persons."
A/HRC/RES/16/11: Human Rights Council resolution of March 24, 2011, "Human rights and the environment."
A/HRC/RES/18/22: Human Rights Council resolution of September 30, 2011, "Human rights and climate change."
A/HRC/RES/19/10: Human Rights Council resolution of March 22, 2012, "Human rights and the environment."
A/HRC/RES/25/21: Human Rights Council resolution of March 28, 2014, "Human rights and the environment."
A/HRC/RES/26/27: Human Rights Council resolution of June 27, 2014, "Human rights and climate change."
A/HRC/RES/28/11: Human Rights Council resolution of March 26, 2015, "Human rights and the environment."
A/HRC/RES/29/15: Human Rights Council resolution of July 2, 2015, "Human rights and climate change."
A/HRC/RES/32/33: Human Rights Council resolution of July 1, 2016, "Human rights and climate change."

A/HRC/RES/6/13: Human Rights Council resolution of September 27, 2008, "The Social Forum."

A/HRC/RES/7/23: Human Rights Council resolution of March 28, 2008, "Human rights and climate change."

A/RES/25/2749: General Assembly resolution of December 12, 1970, "Declaration of principles governing the sea-bed and the ocean floor, and the subsoil thereof, beyond the limits of national jurisdiction."

A/RES/43/53: General Assembly resolution of December 6, 1988, "Protection of global climate for present and future generations of mankind."

A/RES/45/100: General Assembly resolution of December 14, 1990, "Humanitarian assistance to victims of natural disasters and similar emergency situations."

A/RES/46/182: General Assembly resolution of December 19, 1991, "Strengthening of the coordination of humanitarian emergency assistance of the United Nations."

A/RES/48/141: General Assembly resolution of December 20, 1993, "High Commissioner for the promotion and protection of all human rights."

A/RES/53/125: General Assembly resolution of December 9, 1998, "Office of the United Nations High Commissioner for Refugees."

A/RES/58/153: General Assembly resolution of December 22, 2003, "Implementing actions proposed by the United Nations High Commissioner for Refugees to strengthen the capacity of his office to carry out its mandate."

A/RES/60/1: General Assembly resolution of September 16, 2005, "World Summit Outcome document."

A/RES/60/251: General Assembly resolution of April 3, 2006, "Human Rights Council."

A/RES/61/295: General Assembly resolution of September 13, 2007, "United Nations Declaration on the Rights of Indigenous Peoples."

A/RES/62/153: General Assembly resolution of December 18, 2007, "Protection of and assistance to internally displaced persons."

A/RES/63/117: General Assembly resolution of December 10, 2008, "Optional Protocol to the International Covenant on Economic, Social and Cultural Rights."

A/RES/64/162: General Assembly resolution of December 18, 2009, "Protection of and assistance to internally displaced persons."

A/RES/66/138: General Assembly resolution of December 19, 2011, "Optional Protocol to the Convention on the Rights of the Child on a communications procedure."

A/RES/70/1: General Assembly resolution of September 25, 2015, "Transforming our world: The 2030 Agenda for Sustainable Development."

A/RES/70/296: General Assembly resolution of July 25, 2016, "Agreement concerning the relationship between the United Nations and the International Organization for Migration."

A/RES/71/1: General Assembly resolution of September 19, 2016, "New York Declaration for Refugees and Migrants."

African Union Convention Governing the Specific Aspects of Refugee Problems in Africa, adopted on September 10, 1969, by the Assembly of Heads of State and Government, CAB/LEG/24.3. www.achpr.org/files/instruments/refugee-convention/achpr_instr_conv_refug_eng.pdf (accessed May 4, 2019)

AG/RES.2429 (XXXVIII-O/08): Human Rights and climate change in the Americas.

BURSON, Bruce L.: [2013] NZIPT 800413. Immigration and Protection Tribunal, Auckland, New Zealand.

[2014] NZIPT 501370–371. Immigration and Protection Tribunal, Auckland, New Zealand.

[2014] NZIPT 800517–520. Immigration and Protection Tribunal, Auckland, New Zealand.

BVerfGE 3, 383, 399 (1954): Judgment of the Bundesverfassungsgericht [Germany].
BVerwG 7 B 93.02 (2003): Judgment of the Bundesverwaltungsgericht [Germany].
BVerwG 7 C 36.07 (2007): Judgment of the Bundesverwaltungsgericht [Germany].
CCPR/C/21/Rev.1/Add.11 (2001): Human Rights Committee, General Comment No. 29: Article 4 (States of Emergency).
CCPR/C/21/Rev.1/Add.9 (1999): Human Rights Committee, General Comment No. 27: Article 12 (Freedom of Movement).
CCPR/C/35/D/196/1985: Gueye et al. ./. France, Communication No. 196/1985, Human Rights Committee final views of April 3, 1989.
CCPR/C/52D/511/1992: Ilmari Lansman et al. ./. Finland, Communication No. 511/1992, Human Rights Committee opinion of November 8, 1994.
CCPR/C/67/D/666/1995: F. Foin ./. France, Communication No. 666/1995, Human Rights Committee final views of November 9, 1999.
CCPR/C/74/D/919/2000 (2002): M. A. Müller and I. Engelhard ./. Namibia, Communication No. 919/2000, Human Rights Committee final views of March 26, 2002.
CCPR/C/95/D/1457/2006: Poma ./. Peru, Communication No. 1457/2006, Human Rights Committee views of March 27, 2009.
CEDAW: Statement of the CEDAW Committee on Gender and Climate Change, 44th session, New York, July 20 – August 7, 2009. www2.ohchr.org/english/bodies/cedaw/docs/Gender_and_climate_change.pdf (accessed May 22, 2019)
CEDAW/C/2010/47/GC.1: Committee on the Elimination of Discrimination against Women (CEDAW), General recommendation No. 27 on older women and protection of their human rights, October 19, 2010.
CEDAW/C/GC/35: Committee on the Elimination of Discrimination against Women (CEDAW), General recommendation No. 35 on gender-based violence against women, updating general recommendation No. 19, July 14, 2017.
CERD/C/SR.1189: Committee on the Elimination of All Forms of Racial Discrimination, Summary Record of the 1119th [i.e. 1189th] meeting of March 8, 1997.
Chair's summary, entitled "Standing up for humanity: Committing to action," https://interagencystandingcommittee.org/system/files/chairs_summary_standing_up_for_humanity_committing_to_action_-_advance_unedited_version.pdf (accessed June 3, 2019)
COMMISSION ON HUMAN RIGHTS: Decision 1992/110 of February 28, 1992.
Constitution of the International Organization for Migration, Adopted by the Migration Conference in Brussels on December 5, 1951. www.iom.int/constitution (accessed May 24, 2019)
COUNCIL OF EUROPE: Committee of Ministers, Rec(2006)6 of April 5, 2006.
Manual on Human Rights and the Environment, 2nd ed., 2012.
DPI/1294 Rev.2: OHCHR/UNDPI, Vienna Declaration and Programme of Action, 20 Years Working for Your Rights, 1993 World Conference on Human Rights, June 2013.
DPI/2483: Office of the High Commissioner for Human Rights (OHCHR), United Nations Joint Press Kit for the Bali Climate Change Conference, December 3–14, 2007: The Human Rights Impact of Climate Change, November 2007.
E/1991/23: Committee on Economic, Social and Cultural Rights (CESCR), General Comment No. 3 (1990): The nature of States Parties' obligations (Art. 2, para. 1), December 14, 1990.
E/1992/23: Committee on Economic, Social and Cultural Rights (CESCR), General Comment No. 4 (1991): The right to adequate housing (Art. 11 (1)), December 13, 1991.
E/1998/22: Committee on Economic, Social and Cultural Rights (CESCR), General Comment No. 7 (1997): The right to adequate housing (Art. 11 (1) of the Covenant): Forced evictions, May 20, 1997.

E/C.12/1999/5: Committee on Economic, Social and Cultural Rights (CESCR), General Comment No. 12: The right to adequate food (Art. 11), May 12, 1999.
E/C.12/2002/11: Committee on Economic, Social and Cultural Rights (CESCR), General Comment No. 15 (2002): The right to water (Arts. 11 and 12).
E/C.12/2007/1: Committee on Economic, Social and Cultural Rights (CESCR), Statement: An Evaluation of the Obligation to Take Steps to the "Maximum of Available Resources" under an Optional Protocol to the Covenant, 2007.
E/C.12/GC/20: Committee on Economic, Social and Cultural Rights (CESCR), General Comment No. 20 (2009): Non-discrimination in economic, social and cultural rights (Art. 2, para. 2).
E/C.16/2007/2: Report of the Committee of Experts on Public Administration of the United Nations Economic and Social Council on participatory governance and citizens' engagement in policy development, service delivery and budgeting, January 29, 2007.
E/C.19/2008/10: Report of the Special Rapporteurs of the Permanent Forum on Indigenous Issues on the impact of climate change mitigation measures on indigenous peoples and on their territories and lands, March 19, 2008.
E/C.19/2010/7: Study of the Special Rapporteurs of the Permanent Forum on Indigenous Issues on the extent to which climate change policies and projects adhere to the standards set forth in the United Nations Declaration on the Rights of Indigenous Peoples, February 2, 2010.
E/CN.4/1992/23: Analytical report of the Secretary-General on internally displaced persons, February 14, 1992.
E/CN.4/1993/35: Comprehensive study prepared by the Representative of the Secretary-General on the human rights issues related to internally displaced persons, pursuant to Commission on Human Rights resolution 1992/73, January 21, 1993.
E/CN.4/1994/44: Report of the Representative of the Secretary-General on the human rights issues related to internally displaced persons, January 25, 1994.
E/CN.4/1994/44/Add.1: Report of the Representative of the Secretary-General on the human rights issues related to internally displaced persons, Profiles in displacement: Sri Lanka, 25 January 1994.
E/CN.4/1998/53/Add.2: Report of the Representative of the Secretary-General on the human rights issues related to internally displaced persons, Guiding Principles on Internal Displacement, February 11, 1998.
E/CN.4/2006/41/Add.2: Report of the Special Rapporteur on adequate housing as a component of the right to an adequate standard of living, Mission to the Islamic Republic of Iran, March 21, 2006.
E/CN.4/2006/71/Add.1: Report of the Representative of the Secretary-General on the human rights of internally displaced persons, Framework for national responsibility, December 23, 2005.
E/CN.4/RES/1991/44: Commission on Human Rights resolution 1991/44 of March 5, 1991, "Human rights and the environment."
E/CN.4/RES/1993/114: Commission on Human Rights decision 1993/114 of March 10, 1993, "Human rights and the environment."
E/CN.4/RES/2004/55: Commission on Human Rights Resolution 2004/55 of April 20, 2004, "Internally displaced persons."
E/CN.4/RES/2005/25: Commission on Human Rights resolution 2005/25 of April 15, 2005, "Women's equal ownership, access to and control over land and the equal rights to own property and to adequate housing."

E/CN.4/Sub.2/1989/58: Report of the Sub-Commission on Prevention of Discrimination and Protection of Minorities on its 41st session, November 13, 1989.

E/CN.4/Sub.2/1989/SR.18: Sub-Commission on Prevention of Discrimination and Protection of Minorities, Summary record of the 18th meeting, 41st session, August 18, 1989.

E/CN.4/Sub.2/1990/12: Sub-Commission on Prevention of Discrimination and Protection of Minorities, Proposals for a study of the problem of the environment and its relation to human rights: Note, August 3, 1990.

E/CN.4/Sub.2/1991/24: Sub-Commission resolution 1991/24 of August 29, 1991, "Human rights and the environment."

E/CN.4/Sub.2/1991/8: Sub-Commission on Prevention of Discrimination and Protection of Minorities, Preliminary report on human rights and the environment, August 2, 1991.

E/CN.4/Sub.2/1992/7 and Add.1: Sub-Commission on Prevention of Discrimination and Protection of Minorities, Progress report on human rights and the environment, July 2, 1992.

E/CN.4/Sub.2/1992/SR.31: Sub-Commission on Prevention of Discrimination and Protection of Minorities, Summary record of the 31st meeting, August 7, 1992.

E/CN.4/Sub.2/1993/7: Sub-Commission on Prevention of Discrimination and Protection of Minorities, Second progress report on human rights and the environment, August 27, 1992.

E/CN./Sub.2/1993/144: Sub-Commission on Prevention of Discrimination and Protection of Minorities, Decision of March 10, 1993.

E/CN.4/Sub.2/1994/9: Sub-Commission on Prevention of Discrimination and Protection of Minorities, Final report on human rights and the environment, July 6, 1994.

E/CN.4/Sub.2/1997/3: Sub-Commission on Prevention of Discrimination and Protection of Minorities, Note by the Secretariat, Comprehensive examination of thematic issues relating to the elimination of racial discrimination, May 27, 1997.

E/CN.4/Sub.2/1997/7: Report of the Secretary-General on the Expert Seminar on the Practice of Forced Evictions, July 2, 1997.

E/CN.4/Sub.2/2005/17 and Add.1: Sub-Commission on Prevention of Discrimination and Protection of Minorities, Final report of the special rapporteur, Principles on housing and property restitution for refugees and displaced persons, June 28, 2005.

E/CN.4/Sub.2/AC.4/2004/4: Working Group on Indigenous Populations, Preliminary working paper on the principle of free, prior and informed consent of indigenous peoples in relation to development affecting their lands and natural resources that would serve as a framework for the drafting of a legal commentary by the working group on this concept, July 8, 2004.

EC/50/SC/CRP.18: United Nations High Commissioner for Refugees (UNHCR) Executive Committee of June 9, 2000, Complementary Forms of Protection: Their Nature and Relationship to the International Refugee Protection Regime. www.unhcr.org/excom/standcom/3ae68d140/complementary-forms-protection-nature-relationship-international-refugee.html (accessed February 4, 2019)

EC/62/SC/CRP.19: UNHCR Executive Committee, UNHCR's Role in Support of an Enhanced Humanitarian Response for the Protection of Persons Affected by Natural Disasters, June 6, 2011.

Economic and Social Council (ECOSOC) decision 1992/252 of July 29, 1992.

Economic and Social Council (ECOSOC) decision 1991/244, Human rights and the environment (E/1991/86) of 31 May 1991.

EUROPEAN COURT OF HUMAN RIGHTS (ECHR): *Budayeva and Others* v. *Russia*, Appl. No. 15339/02, Judgment of March 20, 2008.

*Öneryıldız* v. *Turkey*, Appl. No. 48939/99, Judgment of November 30, 2004.

EUROPEAN COURT OF JUSTICE (ECJ): *Dieter Janecek* v. *Freistaat Bayern*, Case C-237/07, Judgment of July 25, 2008.
FCCC/ADP/2015/1: Elements for a draft negotiating text, February 25, 2015.
FCCC/ADP/2015/L.6: Draft Paris Agreement, December 5, 2015.
FCCC/AWGLCA/2009/17: Report of the ad hoc working group on long-term cooperative action under the convention on its 8th session, February 5, 2010.
FCCC/AWGLCA/2009/INF.1: Revised negotiating text: Ad-hoc working group on long-term cooperative action under the convention, 6th session, Bonn, June 1–12, 2009: Note by the Secretariat, June 22, 2009.
FCCC/AWGLCA/2009/INF.2: Reordering and consolidation of text in the revised negotiating text: Ad-hoc working group on long-term cooperative action under the convention, 7th session, Bangkok, September 28 – October 9, 2009, and Barcelona, November 2–6, 2009: Note by the Secretariat, September 15, 2009.
FCCC/AWGLCA/2010/14: Negotiating text: Note by the Secretariat, August 13, 2010.
FCCC/AWGLCA/2010/6: Text to facilitate negotiations among parties: Note by the Chair, May 17, 2010.
FCCC/CP/2010/7/Add.1, March 15, 2011, Decision 1/CP.16, The Cancun Agreements: Outcome of the work of the ad hoc working group on long-term cooperative action under the convention.
FCCC/CP/2012/8/Add.1, 28 February 2013, 3/CP.18, Approaches to address loss and damage associated with climate change impacts in developing countries that are particularly vulnerable to the adverse effects of climate change to enhance adaptive capacity.
FCCC/CP/2015/10/Add.1: Report of the Conference of the Parties on its 21st session, held in Paris from November 30 to December 13, 2015, Addendum, Part two: Action taken by the Conference of the Parties at its 21st session, Decision 1/CP.21 Adoption of the Paris Agreement, January 29, 2016.
FCCC/INFORMAL/84: United Nations Framework Convention on Climate Change, 1992.
FCCC/SB/2018/1: Report of the Executive Committee of the Warsaw International Mechanism for Loss and Damage associated with Climate Change Impacts, October 15, 2018.
Geneva Convention Relative to the Protection of Civilian Persons in Time of War, adopted on August 12, 1949, by the Diplomatic Conference for the Establishment of International Conventions for the Protection of Victims of War, held in Geneva from April 21 to August 12, 1949. www.un.org/en/genocideprevention/documents/atrocity-crimes/Doc.33_GC-IV-EN.pdf (accessed March 22, 2019)
Grundgesetz der Bundesrepublik Deutschland: Basic Law for the Federal Republic of Germany in the revised version published in the Federal Law Gazette Part 3, classification number 100–1, as last amended by Article 1 of the Act of December 23, 2014 (Federal Law Gazette 1, p. 2438). www.gesetze-im-internet.de/englisch_gg/englisch_gg.html#p002 (accessed March 22, 2019)
GUTERRES, António: (United Nations High Commissioner for Refugees) Opening Remarks, High Commissioner's Dialogue on Protection Challenges: Understanding and addressing root causes of displacement, Geneva, December 16, 2015. www.unhcr.org/admin/hcspeeches/567139aa9/high-commissioner-dialogue-protection-challenges-understanding-addressing.html (accessed March 12, 2019)
HCR/MINCOMMS/2011/08: United Nations High Commissioner for Refugees (UNHCR), Background Note for the Roundtables, Intergovernmental event at the ministerial level of Member States of the United Nations on the occasion of the 60th anniversary of the 1951

Convention relating to the Status of Refugees and the 50th anniversary of the 1961 Convention on the Reduction of Statelessness (December 7–8, 2011), November 18, 2011.
HR/PUB/06/8: Frequently Asked Questions on a Human Rights-Based Approach to Development Cooperation, Geneva: OHCHR, 2006.
HR/PUB/12/5: Human Rights Indicators: A Guide to Measurement and Implementation, Geneva: OHCHR, 2012.
HRI/GEN/1/Rev.9 (Vol. I): United Nations Human Rights Committee, General Comment No. 18 (1989): Non-discrimination, November 10, 1989.
INTER-AGENCY STANDING COMMITTEE (IASC): Definition of Complex Emergencies, November 30, 1994, Annex I. https://interagencystandingcommittee.org/system/files/legacy_files/WG16_4.pdf (accessed April 4, 2019)
Operational Guidelines on the Protection of Persons in Situations of Natural Disasters, The Brookings–Bern Project on Internal Displacement, January 2011.
INTERGOVERNMENTAL COMMITTEE FOR EUROPEAN MIGRATION (ICEM), Resolution to Establish a Provisional Intergovernmental Committee for the Movement of Migrants from Europe, 1951.
INTERNATIONAL COVENANT ON CIVIL AND POLITICAL RIGHTS (ICCPR), adopted December 16, 1966, G.A. Res. 2200A (XXI), 21 UN GAOR Supp. (No. 16) at 52, A/6316 (1966), 999 UNTS 171.
INTERNATIONAL COVENANT ON ECONOMIC, SOCIAL AND CULTURAL RIGHTS (ICESCR), adopted December 16, 1966, GA Res. 2200A (XXI), 21 UN GAOR Supp. (No. 16) at 49, UN Doc. A/6316 (1966), 993 UNTS 3.
INTERNATIONAL INSTITUTE FOR ENVIRONMENT AND DEVELOPMENT (IIED): Environmental Refugees, London: Earthscan, 1984.
INTERNATIONAL LABOUR ORGANIZATION (ILO): Convention No. 169/1989 Concerning Indigenous and Tribal Peoples
INUIT CIRCUMPOLAR CONFERENCE (ICC): Petition seeking relief from violations resulting from global warming caused by acts and omissions of the United States of America, December 7, 2005. www.inuitcircumpolar.com/uploads/3/0/5/4/30542564/finalpetitionicc.pdf (accessed March 14, 2019)
INTERGOVERNMENTAL PANEL ON CLIMATE CHANGE (IPCC), Principles Governing IPCC Work, approved at the 14th session (Vienna, October 1–3, 1998) on October 1, 1998, amended at the 21st session (Vienna, November 3 and 6–7, 2003), the 25th session (Mauritius, April 26–28, 2006), the 35th session (Geneva, June 6–9, 2012) and the 37th session (Batumi, October 14–18, 2013).
Kyoto Protocol to the United Nations Framework Convention on Climate Change, 1998. http://unfccc.int/resource/docs/convkp/kpeng.pdf (accessed May 27, 2019)
LANDac Conference: Climate Change-Induced Displacement and Migration, Utrecht, Netherlands, July 9–10, 2015. www.landgovernance.org/assets/2015/01/Climate-change-induced-displacement-and-migration.pdf (accessed May 27, 2017)
Malé Declaration on the Human Dimension of Global Climate Change, November 14, 2007. www.ciel.org/Publications/Male_Declaration_Nov07.pdf. (accessed May 27, 2019)
*Maya Indigenous Community of the Toledo District* v. *Belize* [2004], IACHR Case 12.053, Report No. 40/04, OEA/Ser.L/V/II.122, Doc. 5, Rev.1, at 727.
Ministry of Disaster Management and Relief (MoDMR), National Strategy on the Management of Disaster and Climate Induced Internal Displacement (NSMDCIID), September 13, 2015. http://displacementsolutions.org/peninsula-principles-highlighted-in-bangladeshs-national-strategy-for-disaster-and-climate-induced-internal-displacement/ (accessed April 4, 2019)

NANSEN CONFERENCE: Climate Change and Displacement in the 21st Century, Oslo: Nansen Conference, 2011. www.unhcr.org/4ea969729.pdf (accessed April 24, 2019)

NANSEN INITIATIVE: Agenda for the Protection of Cross-Border Displaced Persons in the Context of Disasters and Climate Change, 2 vols., December 2015.

A Protection Agenda for Disaster-Induced Cross-Border Displacement, Concept Note, September 2012. www.regjeringen.no/upload/UD/Vedlegg/Hum/nansen_prinsipper.pdf (accessed April 12, 2019)

National Office of Oceanic and Atmospheric Research, Global Warming and Hurricanes. www.oar.noaa.gov/spotlite/archive/spot_gfdl.html (accessed June 2, 2019)

OFFICE FOR THE COORDINATION OF HUMANITARIAN AFFAIRS (OCHA): Reference Guide, Normative developments on the coordination of humanitarian assistance in the General Assembly and the Economic and Social Council since the adoption of General Assembly resolution 46/182, 2nd edition, 2011.

Peninsula Principles' *Travaux Préparatoires*, August 2013: on file with the author.

Petition to the Inter-American Commission on Human Rights seeking relief from violations resulting from global warming caused by acts and omissions of the United States, December 7, 2015. http://earthjustice.org/sites/default/files/library/legal_docs/summary-of-inuit-petition-to-inter-american-council-on-human-rights.pdf (accessed February 3, 2019) Protocol on the Protection and Assistance to Internally Displaced Persons, International Conference on the Great Lakes Region, 30 November 2006.

Social and Economic Rights Action Center and the Center for Economic and Social Rights ./. Nigeria ("SERAC ./. Nigeria – the Ogoniland Case"), AfCHPR, Communication 155/96 (2002).

United Nations Convention on Access to Information, Public Participation in Decision-Making and Access to Justice in Environmental Matters, June 25, 1998.

UNITED NATIONS HIGH COMMISSIONER FOR REFUGEES (UNHCR): Climate Change and Statelessness: An Overview, to the 6th session of the Ad Hoc Working Group on Long-Term Cooperative Action (AWG-LCA 6) under the UNFCCC, June 1–12, 2009, Bonn, Germany. http://unfccc.int/resource/docs/2009/smsn/igo/048.pdf (accessed February 4, 2019)

Climate Change, Natural Disasters and Human Displacement: A UNHCR Perspective, August 14, 2009. www.unhcr.org/4901e81a4.pdf (accessed May 8, 2019)

Forced Displacement in the Context of Climate Change: Challenges for States under International Law, Submission to the 6th session of the Ad Hoc Working Group on Long-Term Cooperative Action under the Convention, May 20, 2009.

Statute of the Office of the United Nations High Commissioner for Refugees, Annex to General Assembly resolution 428(V) of December 14, 1950.

Summary of Deliberations on Climate Change and Displacement, Bellagio, February 22–25, 2011. www.unhcr.org/4da2b5e19.pdf (accessed April 14, 2019)

The Environment and Climate Change, updated version October 2015. http://unfccc.int/files/adaptation/groups_committees/loss_and_damage_executive_committee/application/pdf/unhcr_the_environment__and_climate_change.pdf (accessed March 20, 2019)

UNITED NATIONS HIGH COMMISSIONER FOR REFUGEES (UNHCR) Executive Committee: Conclusion on Legal Safety Issues in the Context of Voluntary Repatriation of Refugees, October 8, 2004.

UNITED NATIONS HIGH COMMISSIONER FOR REFUGEES (UNHCR)/ Brookings/ Georgetown University: Planned Relocation, Disasters and Climate Change: Consolidating Good Practices and Preparing for the Future, Background Document,

Sanremo Consultation, March 12–14, 2014. www.unhcr.org/protection/environment/53c4d6f99/background-document-planned-relocation-disasters-climate-change-consolidating.html (accessed March 14, 2019)

UNITED NATIONS HIGH COMMISSIONER FOR REFUGEES (UNHCR) and Sri Lanka Human Rights Commission: Land, Housing and Property: Proposals to the Parties for Comprehensively Addressing Land, Housing and Property Rights in the Context of Refugee and IDP Return within and to Sri Lanka, May 2003.

UNISDR: Terminology on Disaster Risk Reduction, 2009. www.unisdr.org/we/inform/terminology (accessed April 4, 2019)

## STATEMENTS/PRESS RELEASES

BACHELET, Michelle: (UN High Commissioner for Human Rights) Statement at the 24th Conference of the Parties to the UN Framework Convention on Climate Change, Global Commission on Adaptation, "Accelerating Action and Support for Adaptation," Katowice, December 12, 2018. www.ohchr.org/en/newsevents/pages/DisplayNews.aspx?NewsID=24011&LangID=E (accessed January 10, 2019)

COP 21: "States' human rights obligations encompass climate change" – UN expert, Paris, December 3, 2015. www.ohchr.org/EN/NewsEvents/Pages/DisplayNews.aspx?NewsID=16836&LangID=E.#sthash.d9XSXpJA.dpuf (accessed April 16, 2019)

*Guardian*: The first climate justice summit: A pie in the face for the global north, April 16, 2014. www.theguardian.com/global-development-professionals-network/2014/apr/16/climate-change-justice-summit (accessed May 5, 2019)

GUTERRES, António: "Climate Action: Mobilizing the World" [as delivered], New York University Stern School of Business, May 30, 2017. www.un.org/sustainabledevelopment/blog/2017/05/secretary-generals-climate-remarks-at-nyu-stern/ (accessed May 9, 2019)

(United Nations High Commissioner for Refugees) Keynote Speech at the Third Symposium on Corporate Social Responsibility and Humanitarian Assistance, Tokyo, November 26, 2007. www.unhcr.org/admin/ADMIN/476132d911.html (accessed April 16, 2019)

(United Nations High Commissioner for Refugees) United Nations Security Council Briefing, "Maintenance of International Peace and Security: New Challenges to International Peace and Security and Conflict Prevention," November 23, 2011, New York. www.unhcr.org/en-us/admin/hcspeeches/4ee21edc9/ (accessed March 23, 2019)

(United Nations Secretary-General) Remarks at the Conclusion of COP 24, Katowice, December 15, 2018. www.un.org/sg/en/content/sg/statement/2018–12-15/secretary-generals-remarks-the-conclusion-of-the-cop24-0 (accessed January 10, 2019)

Inuit leader and activist Sheila Watt-Cloutier's testimony before the Inter-American Commission on Human Rights, March 1, 2007. http://earthjustice.org/sites/default/files/library/legal_docs/testimony-before-iachr-on-global-warming-human-rights-by-sheila-watt-cloutier.pdf (accessed May 4, 2019)

IRIN News: "Global: Taking on the land-grabbers," October 26, 2010. www.irinnews.org/news/2010/10/26/taking-land-grabbers (accessed June 2, 2019)

Joint statement by UN Special Procedures on the occasion of World Environment Day (June 5, 2015), Climate Change and Human Rights. www.ohchr.org/EN/NewsEvents/Pages/DisplayNews.aspx?NewsID=16049&LangID=E#sthash.Ym84pjPp.dpuf (accessed April 16, 2019)

KANG, Kyung-wha (Deputy High Commissioner for Human Rights): Address on "Climate Change and Human Rights," Conference of the Parties to the United Nations Framework

Convention on Climate Change and its Kyoto Protocol, December 3–14, 2007, Bali, Indonesia. http://newsarchive.ohchr.org/EN/NewsEvents/Pages/DisplayNews.aspx?NewsID=200&LangID=E (accessed January 3, 2019)

KHOR, Martin (South Centre): A Human Rights Approach to Climate Change, Statement at the annual full-day discussion on specific themes related to human rights and climate change at the 28th session of the Human Rights Council, March 6, 2015. www.ohchr.org/EN/Issues/HRAndClimateChange/Pages/Discussion6March2015.aspx (accessed April 7, 2019)

LECKIE, Scott: Two Climate Change and Displacement Scenarios: The Choice Is Ours, A Presentation to the Turning Point, 2009, Annual Meeting of the Climate Action Network Australia (CANA), March 11, 2009.

Letter from the Assistant Executive Secretary of the Inter-American Commission on Human Rights (IACHR) of the Organization of American States (OAS) to the Inuit Alliance, Earthjustice and the Centre for International Environmental Law of February 1, 2007. www.ciel.org/Publications/IACHR_Response_1Feb07.pdf (accessed April 16, 2019)

Open Letter from Special Procedures mandate-holders of the Human Rights Council to the State Parties to the UN Framework Convention on Climate Change on the occasion of the meeting of the Ad Hoc Working Group on the Durban Platform for Enhanced Action in Bonn (October 20–25, 2014), October 17, 2014. www.ohchr.org/Documents/HRBodies/SP/SP_To_UNFCCC.pdf (accessed March 20, 2019)

PICTET GROUP CHARITABLE FOUNDATION: Newsletter Impact, No. 1, October 2014. www.group.pictet/corporate/en/home/about_pictet/corporate_citizenship/charitable_foundation/impact_linkedin.html (accessed April 24, 2019)

PR/OHCHR/20March2015: "New framework for Disaster Risk Reduction – UN experts urge move to effective action," Joint statement by Mr. Chaloka Beyani, Special Rapporteur on the human rights of internally displaced persons; Ms. Rosa Kornfeld-Matte, Independent Expert on the enjoyment of all human rights by older persons; Ms. Catalina Devandas Aguilar, Special Rapporteur on the rights of persons with disabilities; and Ms. Victoria Tauli Corpuz; Special Rapporteur on the rights of indigenous peoples. www.ohchr.org/en/NewsEvents/Pages/DisplayNews.aspx?NewsID=15733&LangID=E (accessed March 2, 2019)

PR/OHCHR/20March2015: Statement of the Independent Expert on the enjoyment of all human rights by older persons, Side-event on "Building climate resilience: the rights of groups in focus," March 18, 2015. www.ohchr.org/en/NewsEvents/Pages/NewsSearch.aspx?MID=IE_OlderPersons (accessed May 3, 2019)

PR/OHCHR/20March2015: "Vanuatu: After devastation, UN experts call to protect human rights of all in disaster response," www.ohchr.org/en/NewsEvents/Pages/DisplayNews.aspx?NewsID=15739&LangID=E (accessed March 24, 2019)

PR/OHCHR/6December2018: Joint statement of the United Nations Special Procedures Mandate Holders on the occasion of the 24th Conference of the Parties to the UNFCCC. www.ohchr.org/EN/NewsEvents/Pages/DisplayNews.aspx?NewsID=23982&LangID=E (accessed January 10, 2019)

Prime Minister Bainimarama's Remarks at the Side Event, Commitments and Opportunities to Implement the Nansen Initiative Protection Agenda, Turkey, May 25, 2016. www.fiji.gov.fj/Media-Center/Speeches/HON-PM-BAINIMARAMA-S-REMARKS-AT-THE-SIDE-EVENT–C.aspx (accessed March 20, 2019)

REPRESENTATIVE OF THE SECRETARY-GENERAL ON THE HUMAN RIGHTS OF INTERNALLY DISPLACED PERSONS: Displacement Caused by the Effects of

Climate Change: Who Will Be Affected and What Are the Gaps in the Normative Frameworks for Their Protection?, October 2008.
ROBINSON, Mary: (President, Mary Robinson Foundation – Climate Justice) Statement at the annual full-day discussion on specific themes related to human rights and climate change at the 28th session of the Human Rights Council, March 6, 2015. www.ohchr.org/EN/Issues/HRAndClimateChange/Pages/Discussion6March2015.aspx (accessed April 7, 2019)
Comments at 2nd Interagency Workshop on Implementing a Rights-Based Approach in the Context of UN Reform, May 2003.
Opening of the Office of the High Commissioner for Human Rights (OHCHR), Expert Meeting on Climate Change and Human Rights, Geneva, October 6, 2016. www.ohchr.org/Documents/Issues/ClimateChange/EM2016/RobinsonHighLevelStatement.pdf (accessed January 24, 2019)
Road to Paris, Science for Smart Policies, Q & A with Mary Robinson: What Is Climate Justice?, July 2014 – April 2016. http://roadtoparis.info/2015/07/29/qa-with-mary-robinson-what-is-climate-justice/ (accessed January 24, 2019)
SG/A/1712: Secretary-General Appoints Louise Arbour of Canada Special Representative for International Migration, March 9, 2017. www.un.org/press/en/2017/sga1712.doc.htm (accessed March 20, 2019)
SG/SM/6638: United Nations Press Release of July 14, 1998.
Statement of Dunya Maumoon, State Minister for Foreign Affairs of the Maldives, on March 1, 2012. https://extranet.ohchr.org/sites/hrc/HRCSessions/RegularSessions/19thSession/OralStatements/Maldives2.pdf (accessed April 3, 2019)
Statement of Italy on behalf of the European Union of June 27, 2014, Human Rights and Climate Change, EU Explanation of Position, UN Human Rights Council, 26th Session, item 3, resolution L.33/Rev.1.
Statement of the European Union Delegation at the 22nd session of the Human Rights Council delivered during the interactive dialogue with the Independent Expert on the issue of human rights obligations relating to the enjoyment of a safe, clean, healthy and sustainable environment, March 7, 2013.
The United Nations Special Rapporteurs on the rights of persons with disabilities, Ms. Catalina Devandas Aguilar; the issue of human rights obligations relating to the enjoyment of a safe, clean, healthy, and sustainable environment, Mr. John H. Knox; extreme poverty and human rights, Mr. Philip Alston; the human right to safe drinking water and sanitation, Mr. Léo Heller; and the Independent Expert on human rights and international solidarity, Ms. Virginia Dandan, The Effects of Climate Change on the Full Enjoyment of Human Rights, April 30, 2015. www.thecvf.org/wp-content/uploads/2015/05/humanrightsSRHRE.pdf (accessed March 4, 2019)
TÜRK, Volker: Statement to the 68th session of the Executive Committee of the High Commissioner's Programme, October 5, 2017.
UNITED NATIONS HIGH COMMISSIONER FOR REFUGEES (UNHCR): Statement at the United Nations Conference on Environment and Development (UNCED), Rio de Janeiro, June 10, 1992.
UNITED STATES OF AMERICA: Statement under Agenda Item 3, Clustered Interactive Dialogue with Special Rapporteur on internally displaced persons and the Special Rapporteur on extrajudicial, summary or arbitrary executions, 32nd session of the UN Human Rights Council, June 20, 2016. https://geneva.usmission.gov/2016/06/20/dialogue-with-special-rapporteurs-on-internally-displaced-persons-and-on-extrajudicial-summary-or-arbitrary-executions/ (accessed May 8, 2019)

REPORTS/PUBLICATIONS

BROOKINGS INSTITUTION: From Responsibility to Response: Assessing National Approaches to Internal Displacement, Elizabeth Ferris, Erin Mooney and Chareen Stark, November 1, 2011.

Internal Displacement and the Construction of Peace, Summary Report, Bogota, Colombia, November 11–12, 2008.

Moving beyond Rhetoric: Consultation and Participation with Populations Displaced by Conflict or Natural Disasters, October 2008.

On the Front Line of Climate Change and Displacement: Learning from and with Pacific Island Countries, September 2011.

Protecting Internally Displaced Persons: A Manual for Law and Policymakers, October 2008.

The Year That Shook the Rich: A Review of Natural Disasters in 2011, March 2012.

When Displacement Ends: A Framework for Durable Solutions, June 2007.

CARE, CIESIN, UNHCR, UNU-EHS, World Bank: In Search of Shelter: Mapping the Effects of Climate Change on Human Migration and Displacement, 2009.

CENTRE ON HOUSING RIGHTS AND EVICTIONS (COHRE): Forced Evictions: Violations of Human Rights No. 7, Global Survey on Forced Evictions, Geneva, 1994.

Forced Evictions: Violations of Human Rights No. 10, Global Survey on Forced Evictions, Geneva, 2006.

Sources No. 4: Legal Resources for Housing Rights: International and National Standards, Geneva, 2000.

DISPLACEMENT SOLUTIONS: Activity Reports 2007/2008, 30 March 2009.

Activity Reports 2009/2010, 11 April 2011. http://displacementsolutions.org/category/annual-reports/ (accessed June 4, 2019)

An Overview on the Relocation of Guna Indigenous Communities in Gunayala, Panama, Mission Report, October 2016.

Bridging the HLP Gap, The Need to Effectively Address Housing, Land and Property Rights during Peace Negotiations and in the Context of Refugee/IDP Return: Preliminary Recommendations to the Government of Myanmar, Ethnic Actors and the International Community, prepared by the Director of Displacement Solutions Scott Leckie supported by the Swiss Federal Department of Foreign Affairs (FDFA), Directorate of Political Affairs DP, Human Security Division: Peace, Human Rights, Humanitarian Policy, Migration, June 2013. http://displacementsolutions.org/wp-content/uploads/DIS2383-Bridging-The-Gap-v13.pdf (accessed July 4, 2019)

Building on Results: Climate Change and Displacement Initiative (CCDI), Funding Proposal 2014–2016, Geneva, 2014.

Climate Change Displaced Persons and Housing, Land and Property Rights Preliminary Strategies for Rights-Based Planning and Programming to Resolve Climate Induced Displacement, February 10, 2010.

Climate Displacement and Human Rights: Where to Go from Here?, Displacement Solutions Expert Roundtable Discussion, Centre for Humanitarian Dialogue, Geneva, June 11, 2015. http://displacementsolutions.org/wp-content/uploads/2015/08/2015-06-19-Climate-Displacement-Human-Rights-11-june2015.docx (accessed May 4, 2019)

Climate Displacement and Planned Relocation in Colombia: The Case of Gramalote, August 2015. http://displacementsolutions.org/?s=colombia (accessed April 10, 2019)

Housing, Land and Property Rights and the Quest for Peace: Guidance for Peace Mediators, 20 May 2011. http://displacementsolutions.org/?s=peace (accessed May 4, 2019)

Land Access for Climate Displaced Persons and Communities. Roundtable Hosted by Displacement Solutions and The Global Migration Centre of the Geneva Graduate Institute of International and Development Studies, 2016. http://displacementsolutions.org/wp-content/uploads/2016/06/2016-DS-Land-Access-for-Climate-Displaced-Persons-and-Communities-Roundtable-Report.pdf (accessed May 4, 2019)

Land Solutions for Climate Displacement Works in Colombia: Government-Supported Gramalote Relocation Moves Ahead, March 20, 2017. http://displacementsolutions.org/land-solutions-for-climate-displacement-works-in-colombia-government-supported-gramalote-relocation-moves-ahead/ (accessed May 20, 2019)

One Step at a Time: The Relocation Process of the Gardi Sugdub Community in Gunayala, Panama, 2015. http://displacementsolutions.org/wpcontent/uploads/2015/07/One-Step-at-a-Time-the-Relocation-Process-of-the-Gardi-Sugdub-Community-In-Gunayala-Panama.pdf (accessed May 4, 2019)

Regulatory Obstacles to Rapid and Equitable Emergency and Interim Shelter Solutions after Natural Disasters, 2011.

Roundtable held on April 25–26, 2016 in Geneva, meeting report on Land Access for Climate Displaced Persons and Communities, Roundtable hosted by Displacement Solutions & The Global Migration Centre of the Geneva Graduate Institute of International and Development Studies. http://displacementsolutions.org/wp-content/uploads/2016/06/2016-DS-Land-Access-for-Climate-Displaced-Persons-and-Communities-Roundtable-Report.pdf (accessed May 4, 2019)

The Climate Displacement Land Initiative, Climate Justice for the Climate Displaced, 2016.

The Climate Displacement Land Initiative, Funding Proposal 2016–2020, Climate Justice for the Climate Displaced, 2015.

The Peninsula Principles in Action, Climate Change and Displacement in the Autonomous Region of Gunayala, Panama, Mission Report, Geneva, July 2014. www.nytimes.com/interactive/2014/03/27/world/climate-rising-seas.html?hp&_r=1 (accessed May 4, 2019)

The Rights of Climate Displaced Persons, A Quick Guide, Geneva, 2015.

DISPLACEMENT SOLUTIONS/ APPLIED RESEARCH ASSOCIATION ON JUSTICE, PEACE AND DEVELOPMENT (ARA-JPD)/ OFFICE OF THE HIGH COMMISSIONER FOR HUMAN RIGHTS (OHCHR): Side Event on Building Climate Resilience, The Rights of Groups in Focus, March 18, 2015. www.ohchr.org/Documents/Issues/OlderPersons/ReportSide-Event-Climate-Resilience-recueil.pdf (accessed May 4, 2019)

EARTHJUSTICE: Environmental Rights Report, Human Rights and the Environment, Materials for the 61st session of the United Nations Commission on Human Rights, Geneva, March 14 – April 22, 2005.

Global Warming and Human Rights gets Hearing on the World Stage, Earthjustice Managing Attorney Martin Wagner gives Testimony Before Inter-American Commission on Human Rights. http://earthjustice.org/sites/default/files/library/legal_docs/testimony-before-iachr-on-global-warming-human-rights-by-martin-wagner.pdf (accessed May 4, 2019)

FAO/IDMC/OCHA/OHCHR/UN-Habitat/UNHCR, Handbook, Housing and Property Restitution for Refugees and Displaced Persons, Implementing the "Pinheiro Principles," March 2007.

GOVERNMENT OFFICE FOR SCIENCE: Foresight: Migration and Global Environmental Change, Final Project Report, 2011. https://assets.publishing.service.gov.uk/government/uploads/system/uploads/attachment_data/file/287717/11-1116-migration-and-global-environmental-change.pdf (accessed April 5, 2019)

GREENPEACE: Klimaflüchtlinge, Die verleugnete Katastrophe, 2007.
HANSSON, Michael: (Former President of the Governing Body of the Nansen International Office for Refugees) Report on the Activities of the Office from July 1 to December 31, 1938. [A.19.1939. xii] (1939. xii. B.2); Statutes of Nansen International Office for Refugees as Approved by the Council on January 19, 1931, League of Nations, Official Journal, February 1931.
The Refugee Problem and the League of Nations, Conference held at the Norwegian Nobel Institute, Oslo, Geneva, 1938, Nansen International Office for Refugees, Report by the Governing Body to the Twelfth Assembly of League of Nations [A.27.1931.].
INTERNATIONAL BAR ASSOCIATION: Achieving Justice and Human Rights in an Era of Climate Disruption, Climate Change Justice and Human Rights Task Force Report, 2014.
INTERNATIONAL ORGANIZATION FOR MIGRATION (IOM): International Migration Law, Glossary on Migration, 2004. www.iomvienna.at/sites/default/files/IML_1_EN.pdf. (accessed February 7, 2019)
Migration, Environment and Climate Change: Evidence for Policy (MECLEP), Glossary, July 2014. http://publications.iom.int/system/files/pdf/meclep_glossary_en.pdf?languag e=en. (accessed May 13, 2019)
Taking Sendai Forward, IOM Strategic Work Plan on Disaster Risk Reduction & Resilience 2017–2020, 2013.
INTERNATIONAL PANEL ON CLIMATE CHANGE (IPCC): Climate Change 2001: Impacts, Adaptation and Vulnerability, Contribution of Working Group II to the Third Assessment Report of the Intergovernmental Panel on Climate Change. www.ipcc.ch /ipccreports/tar/wg2/pdf/wg2TARchap19.pdf (accessed May 4, 2019)
Climate Change 2001: Synthesis Report, Contribution of Working Groups I, II, and III to the Third Assessment Report of the Intergovernmental Panel on Climate Change. www .grida.no/publications/267 (accessed May 4, 2019)
Climate Change 2007: Impacts, Adaptation and Vulnerability. Contribution of Working Group II to the Fourth Assessment Report of the Intergovernmental Panel on Climate Change. www.ipcc.ch/pdf/assessment-report/ar4/wg2/ar4_wg2_full_report.pdf (accessed May 13, 2019)
Climate Change 2007: Impacts, Adaptation and Vulnerability, Summary for Policymakers. Part A: Global and Sectoral Aspects, Contribution of Working Group II to the Fourth Assessment Report of the Intergovernmental Panel on Climate Change. www.ipcc.ch/pdf/assessment-report/ar4/wg2/ar4-wg2-spm.pdf (accessed May 13, 2019)
Climate Change 2007: Mitigation. Contribution of Working Group III to the Fourth Assessment Report of the Intergovernmental Panel on Climate Change. www.ipcc.ch /pdf/assessment-report/ar4/wg3/ar4_wg3_full_report.pdf (accessed May 13, 2019)
Climate Change 2007: Mitigation, Summary for Policymakers. Contribution of Working Group III to the Fourth Assessment Report of the Intergovernmental Panel on Climate Change. www.ipcc.ch/pdf/assessment-report/ar4/wg3/ar4-wg3-spm.pdf (accessed May 13, 2019)
Climate Change 2014: Impacts, Adaptation and Vulnerability, Summary for Policymakers. www.ipcc.ch/pdf/assessment-report/ar5/wg2/ar5_wgII_spm_en.pdf (accessed December 5, 2019).
Climate Change 2014: Synthesis Report. Contribution of Working Groups I, II and III to the Fifth Assessment Report of the Intergovernmental Panel on Climate Change. www.ipcc.ch /pdf/assessment-report/ar5/syr/SYR_AR5_FINAL_full_wcover.pdf (accessed May 13, 2019)
First Assessment Report, 1990. www.ipcc.ch/ipccreports/far/wg_I/ipcc_far_wg_I_full_report .pdf (accessed May 4, 2019)

Managing the Risks of Extreme Events and Disasters to Advance Climate Change Adaptation, A Special Report of Working Groups I and II of the Intergovernmental Panel on Climate Change, Summary for Policymakers. Cambridge: Cambridge University Press, 2012.

Policymakers' Summary of the Potential Impacts of Climate Change, Report from Working Group II to IPCC, Intergovernmental Panel on Climate Change, 1990. www.ipcc.ch/ipccreports/far/wg_I/ipcc_far_wg_I_spm.pdf (accessed May 4, 2019)

JÄGER, Jill/ FRÜHMANN, Johannes/ GRÜNBERGER, Sigrid/ VAG, Andras (Eds.): Environmental Change and Forced Migration Scenarios, Synthesis Report, 2009. www.each-for.eu/documents/EACH-FOR_Synthesis_Report_090515.pdf (accessed February 1, 2019)

OFFICE FOR THE COORDINATION OF HUMANITARIAN AFFAIRS (OCHA): Understanding the Climate–Conflict Nexus from a Humanitarian Perspective: A New Quantitative Approach, Occasional Policy Paper, 2016.

OFFICE FOR THE COORDINATION OF HUMANITARIAN AFFAIRS (OCHA)/ INTERNAL DISPLACEMENT MONITORING CENTRE (IDMC)/ NORWEGIAN REFUGEE COUNCIL (NRC): Monitoring Disaster Displacement in the Context of Climate Change, Findings of a study by the United Nations Office for the Coordination of Humanitarian Affairs and the Internal Displacement Monitoring Centre, 2009.

OFFICE OF THE UNITED NATIONS HIGH COMMISSIONER FOR HUMAN RIGHTS (OHCHR): Applying a Human Rights-Based Approach to Climate Change Negotiations, Policies and Measures, Guidance Note, 2010.

Concept Note for the Half-Day of General Discussion on Gender-Related Dimensions of Disaster Risk Reduction and Climate Change. www.ohchr.org/Documents/HRBodies/CEDAW/ClimateChange/ConceptNote.pdf (accessed February 4, 2019)

Indigenous Peoples and the United Nations Human Rights System, Fact Sheet No. 9/Rev.2, New York, Geneva 2013

ORGANISATION FOR ECONOMIC CO-OPERATION AND DEVELOPMENT (OECD): Development Assistance Committee Guidelines on Aid and Environment, No. 3, Guidelines for Aid Agencies on Involuntary Displacement and Resettlement in Development Projects, 1992.

Integrating Human Rights into Development, 2nd ed., Default Book Series, March 2013.

UNEP, in cooperation with International Organization for Migration (IOM), Office for the Coordination of Humanitarian Affairs (OCHA), United Nations University (UNU-EHS), The Permanent Interstate Committee for Drought Control in the Sahel (CILSS): Livelihood Security, Climate Change, Migration and Conflict in the Sahel, Geneva, 2011.

UN-HABITAT: Global Report of Human Settlements: Enhancing Urban Safety and Security, 2007.

UN-HABITAT/UNOHCHR: Forced Evictions, Fact Sheet No. 25, Rev.1, New York, Geneva, 2014.

UNITED NATIONS DEVELOPMENT PROGRAMME (UNDP): Human Development Report 2007/2008.

UNITED NATIONS HIGH COMMISSIONER FOR REFUGEES (UNHCR): Handbook Voluntary Repatriation: International Protection, Geneva: UNHCR, 1996.

Housing, Land and Property Rights in Post-Conflict Societies: Proposals for a New United Nations Institutional and Policy Framework, Scott Leckie UNHCR Consultant, Department of International Protection, Legal and Protection Policy Research Series, PPLA/2005/01, March 2005. www.unhcr.org/protect/PROTECTION/425683e02a5.pdf (accessed January 20, 2019)

Planned Relocation, Disasters and Climate Change: Consolidating Good Practices and Preparing for the Future, Report, Sanremo, Italy, March 12–14, 2014. www.unhcr.org/protection/environment/54082cc69/final-report-planned-relocation-disasters-climate-change-consolidating.html (accessed February 15, 2019)
The State of the World's Refugees 1993. www.unhcr.org/3eeedcf7a.html (accessed March 7, 2019)
UNITED NATIONS OFFICE FOR DISASTER RISK REDUCTION (UNISDR): Global Assessment Report on Disaster Reduction, Risk and Poverty in Changing Climate, 2009.
VERBRUGGEN, Aviel/ MOOMAW, William/ NYBOER, John (Eds.): Annex I: Glossary, Acronyms, Chemical Symbols and Prefixes. In IPCC Special Report on Renewable Energy Sources and Climate Change Mitigation, 2011. www.ipcc.ch/pdf/special-reports/srren/SRREN_Annex_Glossary.pdf (accessed January 17, 2019)
WORLD BANK: Assessing the Impact of Climate Change on Migration and Conflict, n.d.

# Annex

## Final Text
## Peninsula Principles on Climate Displacement within States

PREAMBLE

*Concerned* that events and processes caused or exacerbated by climate change have and will continue to contribute to displacement of populations resulting in the erosion of the rights of those affected, in particular vulnerable and marginalized groups, the loss of assets, housing, land, property and livelihoods, and the further loss of cultural, customary and/or spiritual identity;

*Guided* by the Charter of the United Nations, and REAFFIRMING the Universal Declaration of Human Rights, the International Covenant on Economic, Social and Cultural Rights, the International Covenant on Civil and Political Rights as well as the Vienna Declaration and Program of Action;

*Noting* that these Peninsula Principles build on and contextualize the United Nations Guiding Principles on Internal Displacement to climate displacement within States;

*Understanding* that when an activity raises threats of harm to human health, life or the environment, precautionary measures should be taken;

*Cognisant* that the vast majority of climate displaced persons are not responsible for the processes driving climate change;

*Noting* that while climate displacement can involve both internal and cross-border displacement, most climate displacement will likely occur within State borders;

*Reaffirming* the right of climate displaced persons to remain in their homes and retain connections to the land on which they live for as long as possible, and the need for States to prioritise appropriate mitigation, adaptation and other preventative measures to give effect to that right;

*Reaffirming* further the right of those who may be displaced to move safely and to relocate within their national borders over time;

*Recognising* that voluntary and involuntary relocation often result in the violation of human rights, impoverishment, social fragmentation and other negative consequences, and recognizing the imperative to avoid such outcomes;

*Noting* further that climate displacement if not properly planned for and managed may give rise to tensions and instability within States;

*Acknowledging* that States bear the primary responsibility for their citizens and others living within their jurisdiction, but recognising that, for many States, addressing the issue of and responding to climate displacement presents financial, logistical, political, resource and other difficulties;

*Convinced*, however, that as climate change is a global problem, States should, on request by affected States, provide adequate and appropriate support for mitigation, adaptation, relocation and protection measures, and to provide assistance to climate displaced persons;

*Realising* that the international community has humanitarian, social, cultural, financial and security interests in addressing the problem of climate displacement in a timely, coordinated and targeted manner;

*Realising* further that there has been no significant coordinated response by States to address climate displacement, whether temporary or permanent in nature;

*Recognising* that the United Nations Framework Convention on Climate Change (UNFCCC) and its Kyoto Protocol neither contemplate nor address the issue of climate displacement, and that conferences and meetings of the parties to these instruments have not substantively addressed climate displacement other than in the most general of terms;

*Noting*, however, that paragraph 14(f) of the UNFCCC 16th session of the Convention of the Parties (COP16) Cancun Adaptation Framework refers to enhanced action on adaptation, including "[m]easures to enhance understanding, coordination and cooperation with regard to climate change induced displacement, migration and planned relocation . . .";

*Noting* further that UNFCCC COP18 in Doha decided to establish, at UNFCCC COP19, institutional arrangements to address loss and damage associated with climate change impacts in developing countries that are particularly vulnerable to the adverse effects of climate change as part of the Cancun Adaptation Framework;

*Recognising* the work being undertaken by the United Nations and other intergovernmental and non-governmental agencies to address climate displacement and related factors;

*Realising* the need for a globally applicable normative framework to provide a coherent and principled approach for the collaborative provision of pre-emptive assistance to those who may be displaced by the effects of climate change, as well as effective remedial assistance to those who have been so displaced, and legal protections for both;

*Acknowledging* the IASC Operational Guidelines on the Protection of Persons in Situations of Natural Disasters, the Hyogo Framework for Action, the UN Principles on Housing and Property Restitution for Refugees and Displaced Persons and others, the incorporation of a number of their principles within these Peninsula Principles, and their application to climate displaced persons;

*Acknowledging* also regional initiatives addressing internal displacement such as the African Union Convention for the Protection and Assistance of Internally Displaced Persons in Africa;

*Noting* the work of the Nansen Initiative on disaster-induced cross-border displacement;

*Noting* that these Peninsula Principles, addressing climate displacement within States, necessarily complement other efforts to address cross-border displacement; and

*Recognising* judicial decisions and the writings of eminent jurists and experts as a source of international law, and acknowledging their importance and contribution to formulating the present Peninsula Principles, these Peninsula Principles on Climate Displacement ("Peninsula Principles") provide as follows:

## I INTRODUCTION

### *Principle 1: Scope and Purpose*

These Peninsula Principles:

(a) provide a comprehensive normative framework, based on principles of international law, human rights obligations and good practice, within which the rights of climate displaced persons can be addressed;
(b) address climate displacement within a State and not cross-border climate displacement; and
(c) set out protection and assistance principles, consistent with the UN Guiding Principles on Internal Displacement, to be applied to climate displaced persons.

### *Principle 2: Definitions*

For the purposes of these Peninsula Principles:

(a) "Climate change" means the alteration in the composition of the global atmosphere that is in addition to natural variability over comparable time periods (as defined by the IPCC).
(b) "Climate displacement" means the movement of people within a State due to the effects of climate change, including sudden and slow-onset environmental events and processes, occurring either alone or in combination with other factors.
(c) "Climate displaced persons" means individuals, households or communities who are facing or experiencing climate displacement.
(d) "Relocation" means the voluntary, planned and coordinated movement of climate displaced persons within States to suitable locations, away from risk-prone

areas, where they can enjoy the full spectrum of rights including housing, land and property rights and all other livelihood and related rights.

*Principle 3: Non-Discrimination, Rights and Freedoms*

(a) States shall not discriminate against climate displaced persons on the basis of their potential or actual displacement, and should take steps to repeal unjust or arbitrary laws and laws that otherwise discriminate against, or have a discriminatory effect on, climate displaced persons.
(b) Climate displaced persons shall enjoy, in full equality, the same rights and freedoms under international and domestic law as do other persons in their country, in particular housing, land and property rights.
(c) States should ensure that climate displaced persons are entitled to and supported in claiming and exercising their rights and are provided with effective remedies as well as unimpeded access to the justice system.

*Principle 4: Interpretation*

(a) These Peninsula Principles shall not be interpreted as limiting, altering or otherwise prejudicing rights recognized in international law, including human rights, humanitarian law and related standards, or rights consistent with those laws and standards as recognized under domestic law.
(b) States should interpret these Peninsula Principles broadly, be guided by their humanitarian purpose, and display fairness, reasonableness, generosity and flexibility in their interpretation.

## II GENERAL OBLIGATIONS

*Principle 5: Prevention and Avoidance*

States should, in all circumstances, comply in full with their obligations under international law so as to prevent and avoid conditions that might lead to climate displacement.

*Principle 6: Provision of Adaptation Assistance, Protection and Other Measures*

(a) States should provide adaptation assistance, protection and other measures to ensure that individuals, households and communities can remain in their homes or places of habitual residence for as long as possible in a manner fully consistent with their rights.

(b) States should, in particular, ensure protection against climate displacement and demonstrate sensitivity to those individuals, households and communities within their territory who are particularly dependent on and/or attached to their land, including indigenous people and those reliant on customary rules relating to the use and allocation of land.

### *Principle 7: National Implementation Measures*

(a) States should incorporate climate displacement prevention, assistance and protection provisions as set out in these Peninsula Principles into domestic law and policies, prioritising the prevention of displacement.
(b) States should immediately establish and provide adequate resources for equitable, timely, independent and transparent procedures, institutions and mechanisms – at all levels of government (local, state and national) – to implement these Peninsula Principles and give effect to their provisions through specially earmarked budgetary allocations and other resources to facilitate that implementation.
(c) States should ensure that durable solutions to climate displacement are adequately addressed by legislation and other administrative measures.
(d) States should ensure the right of all individuals, households and communities to adequate, timely and effective participation in all stages of policy development and implementation of these Peninsula Principles, ensuring in particular such participation by indigenous peoples, women, the elderly, minorities, persons with disabilities, children, those living in poverty, and marginalized groups and people.
(e) All relevant legislation must be fully consistent with human rights laws and must in particular explicitly protect the rights of indigenous peoples, women, the elderly, minorities, persons with disabilities, children, those living in poverty, and marginalized groups and people.

### *Principle 8: International Cooperation and Assistance*

(a) Climate displacement is a matter of global responsibility, and States should cooperate in the provision of adaptation assistance (to the maximum of their available resources) and protection for climate displaced persons.
(b) In fulfilling their obligations to prevent and respond to climate displacement within their territory, States have the right to seek cooperation and assistance from other States and relevant international agencies.
(c) States and relevant international agencies, either separately or together, should provide such cooperation and assistance to requesting States, in particular where the requesting State is unable to adequately prevent and respond to climate displacement.

(d) States that are otherwise unable to adequately prevent and respond to climate displacement should accept appropriate assistance and support from other States and relevant international agencies, whether made individually or collectively.

## III CLIMATE DISPLACEMENT PREPARATION AND PLANNING

### *Principle 9: Climate Displacement Risk Management*

States, in terms of climate displacement risk management, monitoring and modeling, using a rights-based approach, should:

(a) identify, design and implement risk management strategies, including risk reduction, risk transfer and risk sharing mechanisms, in relation to climate displacement;
(b) undertake systematic observation and monitoring of, and disaggregated data collection at the household, local, regional and national levels on, current and anticipated climate displacement;
(c) enhance sharing, access to and the use of such data at the household, local, regional and national levels, mindful of the need for data protection and predetermined use of data, facilitate the assessment and management of climate displacement;
(d) model likely climate displacement scenarios (including timeframes and financial implications), locations threatened by climate change, and possible relocation sites for climate displaced persons;
(e) integrate relocation rights, procedures and mechanisms, as defined in these Peninsula Principles, within national laws and policies; and
(f) develop institutional frameworks, procedures and mechanisms with the participation of individuals, households and communities that:
   (i) identify indicators that will, with as much precision as possible, classify where, at what point in time, and relevant to whom, relocation will be required as a means of providing durable solutions to those affected;
   (ii) require and facilitate governmental technical assistance and funding; and
   (iii) outline steps individuals, households and communities can take prior to climate displacement in order to receive such technical assistance and financial support.

### *Principle 10: Participation and Consent*

To enable successful preparation and planning for climate displacement, States should:

(a) ensure that priority consideration is given to requests from individuals, households and communities for relocation;

(b) ensure that no relocation shall take place unless individuals, households and communities (both displaced and host) provide full and informed consent for such relocation;
(c) only require relocation to take place without such consent in exceptional circumstances when necessary to protect public health and safety or when individuals, households and communities face imminent loss of life or limb;
(d) adopt measures that promote livelihoods, acquisition of new skills, and economic prosperity for both displaced and host communities;
(e) make certain that:
  (i) affected individuals, households and communities (both displaced and host) are fully informed and can actively participate in relevant decisions and the implementation of those decisions, including the planning and implementation of laws, policies and programs designed to ensure respect for and protection of housing, land and property rights;
  (ii) basic services, adequate and affordable housing, education and access to livelihoods (without discrimination) will be available for climate displaced persons in the host community at a standard ensuring equity between the host and relocating communities, and consistent with the basic human rights of each;
  (iii) adequate mechanisms, safeguards and remedies are in place to prevent and resolve conflicts over land and resources; and
  (iv) the rights of individuals, households and communities are protected at all stages of the relocation process;
(f) prior to any relocation, prepare a master relocation plan that addresses critical matters including:
  (i) land acquisition;
  (ii) community preferences;
  (iii) transitional shelter and permanent housing;
  (iv) the preservation of existing social and cultural institutions and places of climate displaced persons;
  (v) access to public services;
  (vi) support needed during the transitional period;
  (vii) family and community cohesion;
  (viii) concerns of the host community;
  (ix) monitoring mechanisms; and
  (x) grievance procedures and effective remedies.

*Principle 11: Land Identification, Habitability and Use*

(a) Recognising the importance of land in the resolution of climate displacement, States should:

(i) identify, acquire and reserve sufficient, suitable, habitable and appropriate public and other land to provide viable and affordable land-based solutions to climate displacement, including through a National Climate Land Bank;
(ii) develop fair and just land acquisition and compensation processes and appropriate land allocation programmes, with priority given to those most in need; and
(iii) plan for and develop relocation sites including new human settlements on land not at risk from the effects of climate change or other natural or human hazards and, in so planning, consider the safety and environmental integrity of the new site(s), and ensure that the rights of both those relocated and the communities that host them are upheld.

(b) In order to determine the habitability and feasibility of any relocation site, and to ensure that climate displaced persons being relocated and the relevant jurisdictional authority are in agreement as to the habitability of any such site, States should create and make publicly available specific, geographically appropriate, standard criteria including:
(i) current and future land use;
(ii) restrictions (including those of a customary nature or not otherwise formally codified) associated with the land and its use;
(iii) habitability of the land, including issues such as accessibility, availability of water, vulnerability to climate or other natural or human hazards, and use; and
(iv) feasibility of subsistence/agricultural use, together with mechanisms for climate displaced persons to decide to where they wish to voluntarily relocate.

(c) States should provide easily accessible information to individuals, households and communities concerning:
(i) the nature and extent of the actual and potential changes to the habitability of their homes, lands and places of habitual residence on which they dwell or subsist, resulting from climate change, including the evidence on which such assessments are made;
(ii) evidence that all viable alternatives to relocation have been considered, including mitigation and adaptation measures that could be taken to enable people to remain in their homes and places of habitual residence;
(iii) planned efforts to assist climate displaced persons in relocation;
(iv) available compensation and alternative relocation options if the relocation site offered is unacceptable to climate displaced persons; and
(v) rights under international and domestic law, in particular housing, land and property rights.

(d) States should include in relocation planning:

(i) measures to compensate climate displaced persons for lost housing, land and property;
(ii) assurances that housing, land, property and livelihood rights will be met for all climate displaced persons, including those who have informal land rights, customary land rights, occupancy rights or rights of customary usage, and assurances that such rights are ongoing; and
(iii) assurances that rights to access traditional lands and waters (for example, for hunting, grazing, fishing and religious purposes) are maintained or similarly replicated.

### *Principle 12: Loss and Damage*

States should develop appropriate laws and policies for loss suffered and damage incurred in the context of climate displacement.

### *Principle 13: Institutional Frameworks to Support and Facilitate the Provision of Assistance and Protection*

(a) States should strengthen national capacities and capabilities to identify and address the protection and assistance needs of climate displaced persons through the establishment of effective institutional frameworks and the inclusion of climate displacement in National Adaptation Programmes of Action as appropriate.
(b) States should take all appropriate administrative, legislative and judicial measures, including the creation of adequately funded Ministries, departments, offices and/or agencies at the local (in particular), regional and national levels empowered to develop, establish and implement an institutional framework to:
   (i) enable government technical assistance and funding to prevent, prepare for and respond to climate displacement;
   (ii) support and facilitate the provision of assistance and protection to climate displaced persons;
   (iii) exchange information and cooperate with indigenous peoples, women, the elderly, minorities, persons with disabilities, children, those living in poverty, and marginalized groups.
   (iv) represent the needs of climate displaced persons.
(c) Responsibility for establishing Ministries, departments, offices and/or agencies should lie with national governments, and such governments should consult and collaborate with regional and local authorities, and integrate such Ministries, departments, offices and/or agencies in relevant institutional frameworks.

(d) States should ensure the provision of adequate resources (including points of contact and assistance) at all levels of government that directly address the concerns of climate displaced persons.

## IV DISPLACEMENT

### *Principle 14: State Assistance to Those Climate Displaced Persons Experiencing Displacement but Who Have Not Been Relocated*

(a) States have the primary obligation to provide all necessary legal, economic, social and other forms of protection and assistance to those climate displaced persons experiencing displacement but who have not been relocated.
(b) Protection and assistance activities undertaken by States should be carried out in a manner that respects both the cultural sensitivities prevailing in the affected area and the principles of maintaining family and community cohesion.
(c) States should provide climate displaced persons experiencing displacement but who have not been relocated with a practicable level of age and gender-sensitive humanitarian assistance including, without limitation, as the context requires:
  (i) emergency humanitarian services;
  (ii) evacuation and temporary and effective permanent relocation;
  (iii) medical assistance and other health services;
  (iv) shelter;
  (v) food;
  (vi) potable water;
  (vii) sanitation;
  (viii) measures necessary for social and economic inclusion including, without limitation, anti-poverty measures, free and compulsory education, training and skills development, and work and livelihood options, and issuance and replacement of lost personal documentation; and
  (ix) facilitation of family reunion.

### *Principle 15: Housing and Livelihood*

(a) States should respect, protect and fulfill the right to adequate housing of climate displaced persons experiencing displacement but who have not been relocated, which includes accessibility, affordability, habitability, security of tenure, cultural adequacy, suitability of location, and non-discriminatory access to basic services (for example, health and education).
(b) Where climate displacement results in the inability of climate displaced persons to return to previous sources of livelihood, appropriate measures should be taken to ensure such livelihoods can be continued in a sustainable manner and

will not result in further displacement, and opportunities created by such measures should be available without discrimination of any kind.

*Principle 16: Remedies and Compensation*

Climate displaced persons experiencing displacement but who have not been relocated and whose rights have been violated shall have fair and equitable access to appropriate remedies and compensation.

## V POST-DISPLACEMENT AND RETURN

*Principle17: Framework for Return*

(a) States should develop a framework for the process of return in the event that displacement is temporary and return to homes, lands or places of habitual residence is possible and agreed to by those affected.
(b) States should allow climate displaced persons experiencing displacement to voluntarily return to their former homes, lands or places of habitual residence, and should facilitate their effective return in safety and with dignity, in circumstances where such homes, lands or places of habitual residence are habitable and where return does not pose significant risk to life or livelihood.
(c) States should enable climate displaced persons to decide on whether to return to their homes, lands or places of habitual residence, and provide such persons with complete, objective, up- to-date and accurate information (including on physical, material and legal safety issues) necessary to exercise their right to freedom of movement and to choose their residence.
(d) States should provide transitional assistance to individuals, households and communities during the process of return until livelihoods and access to services are restored.

## VI IMPLEMENTATION

*Principle 18: Implementation and Dissemination*

States, who have the primary obligation to ensure the full enjoyment of the rights of all climate displaced persons within their territory, should implement and disseminate these Peninsula Principles without delay and cooperate closely with inter-governmental organisations, non-government organisations, practitioners, civil society, and community-based groups toward this end.

Adopted by a group of eminent jurists, text writers, legal scholars and climate change experts in Red Hill on the Mornington Peninsula, Victoria, Australia on 18 August 2013.

# Meeting Draft
# The Peninsula Principles on Climate Displacement

## Table of Contents

## PREAMBLE

ACKNOWLEDGING that events and processes caused or exacerbated by climate change have and will continue to contribute to displacement of populations resulting in the erosion of the rights of those affected, in particular vulnerable and marginalized groups, and the loss of assets, in particular losses relating to housing, land, property and livelihoods;

UNDERSTANDING that, when an activity raises threats of harm to human health, life or the environment, precautionary measures should be taken even if some cause-and-effect relationships are not fully established scientifically;

COGNISANT that the vast majority of those displaced or facing displacement have not been participant in, nor are they responsible for, the processes driving climate change;

NOTING that while climate displacement can involve both internal and cross-border displacement, and that most climate displacement will likely occur primarily within State borders;

REAFFIRMING the frequently strong wish of those who may be displaced by the effects of climate change to remain in their homes and retain connections to the land on which they live for as long as possible and, where necessary, to move safely and to relocate within their national borders over time;

AWARE that climate displacement will likely be involuntary and in many instances be permanent in nature, and be caused by a combination of extreme weather events and slow onset environmental change;

NOTING, further, that climate displacement will likely give rise to conflict and intensify existing conflicts;

ACKNOWLEDGING that States bear the primary responsibility for their citizens and others living within their jurisdiction;

CONVINCED, however, that climate change is a global problem, and that the international community has an obligation to assist in resolving that problem through adequate and appropriate mitigation, adaptation, relocation and protection measures, and to provide assistance to climate displaced persons;

REALISING that the international community has humanitarian, social, cultural, financial and security interests in addressing the problem of climate displacement in a timely, coordinated and targeted fashion;

RECOGNISING that, for many States, addressing the issue of and responding to climate displacement presents financial, logistical, political, resourcing and other difficulties, and that there has been no significant coordinated response by States to address climate displacement, whether temporary or permanent in nature;

RECOGNISING that the United Nations Framework Convention on Climate Change (UNFCCC) and its Kyoto Protocol neither contemplate nor address the issue of climate displacement, and that conferences and meetings of the parties to these instruments have failed to substantively address climate displacement other than in the most general of terms;

NOTING, however, that paragraph 14(f) of the UNFCCC 16th session of the Convention of the Parties (COP16) Cancun Adaptation Framework refers to enhanced action on adaptation, including "[m]easures to enhance understanding, coordination and cooperation with regard to climate change induced displacement, migration and planned relocation . . . ";

NOTING, further, that UNFCCC COP18 in Doha decided to establish, at UNFCCC COP19, institutional arrangements to address loss and damage associated with climate change impacts in developing countries that are particularly vulnerable to the adverse effects of climate change as part of the Cancun Adaptation Framework;

RECOGNISING the work being undertaken by numerous United Nations and other intergovernmental agencies to address climate displacement;

REALISING the need for a globally applicable normative framework to provide principles for the collaborative provision of pre-emptive assistance to those who may be displaced by the effects of climate change, as well as effective remedial assistance to those who have been so displaced, and legal protections for both;

ACKNOWLEDGING the value of normative frameworks such as the Guiding Principles on Internal Displacement (the Guiding Principles), the IASC Operational Guidelines on the Protection of Persons in Situations of Natural Disasters, the Hyogo Framework for Action, the UN Principles on Housing and Property Restitution for Refugees and Displaced Persons and others, the incorporation of a number of their principles within these Principles, and their application to climate displaced persons (as defined in these Principles);

ACKNOWLEDGING also regional initiatives addressing internal displacement such as the African Union Convention for the Protection and Assistance of Internally Displaced Persons in Africa (the Kampala Convention);

NOTING the work of the Nansen Initiative on disaster-induced cross-border displacement and its "protection agenda" for people displaced across borders due to natural disasters and the adverse effects of climate change;

NOTING FURTHER that these Peninsula Principles in contrast and complementary to the Nansen Initiative, address climate displacement within States;

RECOGNISING that displacement and involuntary relocation have often caused impoverishment and social fragmentation, and DESIROUS of preventing such consequences as a result of climate displacement;

RECOGNISING the writings of eminent jurists as a source of international law, and acknowledging the importance of such writings in drafting the present Principles; and

ACKNOWLEDGING comments made on earlier drafts of these Peninsula Principles by eminent jurists, text writers, legal scholars and climate change experts from many jurisdictions,

these Peninsula Principles on Climate Displacement ("Principles") provide as follows:

## I INTRODUCTION

### *Principle 1: Purpose*

These Principles:

(a) provide a comprehensive, normative framework, based on principles of international law, existing human rights obligations and best practice, within which the rights of climate displaced persons can be addressed;
(b) set out protection and assistance Principles to be applied to climate displaced persons; and
(c) address climate displacement within a State and not cross-border climate displacement.

### *Principle 2: Definitions of Climate Displacement, Climate Displaced Persons, and Relocation*

In these Principles:

(a) "Climate displacement" means the involuntary movement within a State of individuals, households and communities in urban and/or rural areas due to the effects of climate change (which may include rising sea levels, slow-onset environmental changes, flooding, drought, extreme weather events and other natural disasters).
(b) "Climate displaced persons" means individuals, households or communities in urban and/or rural areas who are threatened with climate displacement or displaced by the effects of climate change, (either alone or in combination with other factors due to, *inter alia*, rising sea levels, slow-onset environmental changes, flooding, drought, extreme weather events and other natural disasters), and whose homes or places of former habitual residence have been rendered temporarily or permanently uninhabitable.
(c) "Relocation" means the voluntary, planned and coordinated movement of climate displaced persons within States, whether individuals, households or communities, to suitable locations away from vulnerable risk-prone areas where they can enjoy the full spectrum of rights including housing, land and property rights and all other livelihood and related rights.

### *Principle 3: Rights, Freedoms and Non-Discrimination*

(a) States shall not discriminate against climate displaced persons on the basis of their displacement or race; colour; gender; sexual orientation; gender identity; language; religion; belief; political or other opinion; national, ethnic or social origin; legal or social status; caste; age; disability; property ownership; place of birth; or any other status.

(b) Climate displaced persons experiencing climate displacement shall enjoy, in full equality, the same rights and freedoms under international and domestic law as do other persons in their country, in particular housing, land and property rights.

### *Principle 4: Interpretation*

(a) These Principles shall not be interpreted as limiting, altering or otherwise prejudicing rights recognized in international law, including human rights, refugee and humanitarian law and related standards, or rights consistent with those laws and standards as recognized under domestic law.
(b) In determining who is a "climate displaced person" (as defined in Principle 2), States should interpret this term broadly, be guided by the humanitarian purpose of these Principles, create appropriate mechanisms to make such determination, and display fairness, reasonableness, generosity and flexibility in this regard.

## II PRINCIPLES OF RESPONSIBILITY

### *Principle 5: Provision of Adaptation Assistance and Protection to Prevent and Address Climate Displacement*

(a) States should provide adaptation assistance and protection to climate displaced persons to ensure that such persons can remain in their homes for as long as possible or can relocate within their State in a manner fully consistent with their rights.
(b) States should, in particular:
  (i) afford protection against climate displacement to those persons and groups under their jurisdiction who are particularly dependent on and/or attached to their land, including indigenous people and those reliant on customary rules relating to the use and allocation of land; and
  (ii) demonstrate sensitivity regarding the importance of their land to such persons and groups.

### *Principle 6: National Implementation Measures*

(a) With the participation of individuals, households or communities potentially displaced by climate change, States should immediately establish and provide adequate resources for equitable, timely, independent and transparent procedures, institutions and mechanisms – at all levels of government (local, state and national) – to implement these Principles and give effect to their provisions.

(b) In particular, States should incorporate climate displacement assistance and protection provisions as set out in these Principles into domestic law and policies, priortise prevention of displacement, and provide adequate resources for their implementation.
(c) States should ensure that all manifestations of, and durable solutions to, climate displacement are adequately addressed by legislation.
(d) All relevant legislation must be fully consistent with relevant human rights laws, and must in particular explicitly protect the rights of vulnerable groups (including women, the elderly, minorities, persons with disabilities, children, indigenous populations, and those living in poverty).
(e) Where existing procedures, institutions and mechanisms can effectively implement these Principles and give effect to their provisions, States should make available adequate and specially earmarked budgetary allocations and other resources to facilitate that implementation.

*Principle 7: International Cooperation and Assistance*

(a) Climate displacement is a matter of global responsibility, and States should be obliged to cooperate in the provision of adaptation assistance (to the maximum of their available resources) and protection to climate displaced persons.
(b) In fulfilling their obligations to climate displaced persons in their territories, States have the right to seek and receive assistance.
(c) A State should, if it is unable (for example, due to lack of resources) to provide assistance to climate displaced persons, seek cooperation and assistance from other States and relevant international agencies, and such other States and agencies, either separately or together, should provide such cooperation and assistance to that State.
(d) States should cooperate in assisting and protecting the rights of climate displaced persons, invoking, as appropriate the "responsibility to protect" doctrine.
(e) Affected States should consider in good faith offers of assistance and support from other States and from humanitarian and like organisations, whether made individually or collectively, in relation to climate displacement.

## III CLIMATE DISPLACEMENT PREPARATION AND PLANNING

*Principle 8: Climate Displacement Risk Management*

States should, in terms of climate displacement risk management, monitoring and modeling:

(a) identify, design and implement risk management strategies, including risk reduction, risk-transfer and risk-sharing mechanisms, in relation to climate displacement;

(b) undertake systematic observation and monitoring of, and disaggregated data collection at the household, local, regional and national levels on, current and anticipated climate displacement;
(c) enhance sharing, access to and the use of such data at the local, regional and national levels to facilitate the assessment and management of climate displacement;
(d) model likely climate displacement scenarios (including timeframes and financial implications), locations threatened by climate change, and possible relocation sites for climate displaced persons;
(e) integrate relocation rights, procedures and mechanisms, as defined in these Principles, within national laws and policies; and
(f) develop institutional frameworks that:
    (i) identify indicators that will, with as much precision as possible, classify where, at what point in time, and relevant to whom, planned relocation will be required as a means of providing durable solutions to those affected;
    (ii) require and facilitate and governmental technical assistance and funding; and
    (iii) outline steps individuals, households and communities must take prior to climate displacement in order to receive such technical assistance and financial support.

*Principle 9: Participation and Consultation*

To enable successful preparation and planning for climate displacement, States should:

(a) ensure that no relocation shall take place unless individuals, households and communities (both displaced and host) provide full and informed consent for such relocation, but for exceptional circumstances where such individuals, households and communities face imminent loss of life or limb;
(b) subject to the consent of individuals, households and communities affected, in planning for climate displacement, include measures that promote livelihoods and economic prosperity and minimize risks in relation to both displaced and host communities;
(c) make certain that:
    (i) affected individuals, households and communities (both displaced and host) are fully informed and can actively participate in relevant decisions and the implementation of those decisions, including the planning and implementation of laws, policies and programs designed to ensure respect for and protection of housing, land and property rights;

(ii) basic services, adequate and affordable housing, education and access to livelihoods (without discrimination) are available for climate displaced persons in the host community at a standard at least equivalent to that enjoyed by that host community;
(iii) adequate mechanisms and safeguards are in place to prevent and resolve conflicts over land and resources; and
(iv) the rights of climate displaced persons are protected at all stages of the relocation process; and

(d) prior to any relocation, prepare a master relocation plan that addresses critical matters including:
(i) land acquisition;
(ii) community preferences;
(iii) transitional shelter and permanent housing;
(iv) the preservation of existing social and cultural institutions of climate displaced persons;
(v) access to public services;
(vi) support needed during the transitional period;
(vii) family and community cohesion;
(viii) concerns of the host community;
(ix) monitoring mechanisms; and
(x) grievance procedures.

### *Principle 10: Land Identification, Habitability and Use*

(a) Recognising the importance of land in the resolution of the climate displacement problem in all of its forms, States should, in the course of planning for and anticipating climate displacement and the relocation of communities:
(i) identify, acquire and reserve sufficient suitable and appropriate public and other land as may be required to provide viable and affordable land-based solutions to all climate displaced persons, including through a national climate land bank;
(ii) develop fair and just land acquisition and compensation processes and appropriate land allocation programmes, with priority given to those most in need;
(iii) plan for and develop new settlements and towns on land not at risk from the effects of climate change or other natural or human hazards and, in so planning, consider the safety and environmental integrity of the new site(s) and ensure that the rights of both those relocated and the communities that host them are upheld; and
(iv) take measures to facilitate the identification of land for when people need to be temporarily or permanently moved as a result of the effects of climate change.

(b) In order to determine the habitability and feasibility of any relocation site, and to ensure that climate displaced persons being relocated and the relevant jurisdictional authority are in agreement as to the habitability of any such site, States should create specific, geographically relevant, standard criteria including:
   (i) current land use;
   (ii) restrictions (including those of a customary nature or not otherwise formally codified) associated with the land and its use;
   (iii) habitability of the land, including issues such as accessibility, availability of water, vulnerability to climate or other natural or human hazards, and use; and
   (iv) feasibility of subsistence/agricultural use, together with mechanisms for communities to decide to where they wish to voluntarily relocate.

(c) States agree that climate displaced persons have the right to participate fully in decisions with regard to their relocation (for example, whether they relocate, destination, timing, the manner in which the movement is to be effected, housing options, livelihood and economic issues) and to be provided with easily accessible information concerning:
   (i) the nature and extent of the changes to their land and habitats resulting from climate change, including the evidence on which such assessments are made;
   (ii) evidence that all other viable alternatives to relocation have been considered (including mitigation and adaptation measures that could be taken to enable people to remain in their communities);
   (iii) planned efforts to assist affected individuals, households and communities in relocating, both short and long-term;
   (iv) available compensation and alternative relocation options if the relocation site offered is unacceptable to the household or relevant community; and
   (v) rights of climate displaced persons under international and domestic law.

(d) States should include in relocation planning:
   (i) measures to compensate climate displaced persons for lost housing, land and property; and
   (ii) assurances that housing, land, property and livelihood rights are met for all climate displaced persons, including those who have informal land rights, customary land rights, occupancy rights or rights of customary usage, and assurances that such rights are ongoing.

(e) States should consider the necessity of amending national laws to ensure that the rights of minority or other vulnerable groups (including women, the elderly, persons with disabilities, children, indigenous populations, and those living in poverty) are fully respected.

*Principle 11: Loss and Damage*

States should develop appropriate laws and policies for loss suffered and damage incurred by climate displaced persons and, more particularly, should:

(a) assess the risk of loss and damage associated with climate displacement in their territory;
(b) identify and develop appropriate approaches to address loss and damage associated with potential and actual climate displacement, including through risk reduction;
(c) undertake systematic observation and monitoring of, and disaggregated data collection at the local, regional and national levels on, the impacts of climate change, and account for likely and actual losses, as appropriate;
(d) enhance understanding of how loss and damage affects those segments of the population already vulnerable to displacement;
(e) examine how approaches to address loss and damage associated with climate displacement may be integrated into climate-resilient development processes; and
(f) involve communities and populations vulnerable to displacement, civil society, the private sector and other relevant stakeholders in the assessment of and response to loss and damage.

*Principle 12: Institutional Framework to Support and Facilitate the Provision of Assistance and Protection*

(a) States should strengthen national capacities and capabilities to identify and address the protection and assistance needs of climate displaced persons through the establishment of appropriate institutional frameworks and the inclusion of climate displacement in National Adaptation Programmes of Action.
(b) States should take all appropriate administrative, legislative and judicial measures, including the creation of adequately funded Ministries, departments, offices and/or agencies at the local (in particular), regional and national levels empowered to develop, establish and implement an institutional framework to:
    (i) enable government technical assistance and funding; outline the steps a community must take prior to displacement in order to receive such technical assistance; and ensure that community is assisted to take those steps;
    (ii) support and facilitate the provision of assistance and protection to climate displaced persons;
    (iii) exchange information and cooperate with civil society, indigenous groups, international organisations and other relevant actors; and
    (iv) represent the needs of potential or actual climate displaced persons.

(c) Responsibility for establishing such departments, offices and/or agencies should lie with national governments, and such governments should consult and collaborate with state, regional and local governments (or authorities), as appropriate, in this regard, and integrate such departments and agencies in such institutional framework.
(d) States should ensure the provision of adequate resources (including points of contact and assistance) at the local level of government which most directly addresses the concerns of climate displaced persons.

### *Principle 13: Bilateral and Regional Arrangements between States*

States should consider entering into bilateral and regional arrangements with other States as a coordinated means of preventing and planning for climate displacement.

### *Principle 14: Prevention and Avoidance*

States should, in all circumstances, ensure respect for their obligations under international law so as to help prevent and avoid conditions that might lead to displacement of persons due to the effects of climate change.

## IV DISPLACEMENT

### *Principle 15: State Assistance to Climate Displaced Persons*

In the event that climate displacement can be neither prevented nor resolved through mitigation, adaptation and relocation, and no other durable solutions are immediately available, and displacement occurs, States should provide climate displaced persons under their jurisdiction with a practicable level of age and gender-sensitive humanitarian assistance including, without limitation:

(a) emergency humanitarian services;
(b) evacuation and temporary and effective permanent relocation;
(c) medical assistance and other health services;
(d) shelter;
(e) food;
(f) clean water;
(g) sanitation;
(h) measures necessary for social and economic inclusion including, without limitation, anti-poverty measures, free and compulsory education, training and skills development; and
(i) facilitation of family reunion.

*Principle 16: Legal, Economic, Social and Other Protection Measures*

(a) States have the primary duty and responsibility to provide all necessary legal, economic, social and other forms of protection and assistance to those climate displaced persons displaced within their borders.
(b) States are responsible for ensuring that their actions in terms of protection and assistance are consistent with both domestic and international law.
(c) Protection and assistance activities undertaken by States should be carried out in a manner that respects both the cultural sensitivities prevailing in the affected area and the principles of maintaining family and community cohesion.

*Principle 17: Prohibition of Arbitrary and Discriminatory Laws*

(a) States should neither adopt nor apply laws that prejudice the rights, freedoms and needs of climate displaced persons.
(b) States should take steps to repeal unjust or arbitrary laws and laws that otherwise discriminate against, or have a discriminatory effect on, climate displaced persons.

*Principle 18: Remedies and Compensation*

Climate displaced persons should have full access to just remedies and compensation and, in securing such remedies and compensation, States should:

(a) ensure that climate displaced persons are entitled to and should be supported in claiming and exercising their rights and are provided with effective remedies as well as unimpeded access to the justice system, including in case of violations of those rights or when conflicts emerge within or between communities affected;
(b) establish legal frameworks to provide (as appropriate) compensation, reparation or restitution on just and satisfactory terms to climate displaced persons; and
(c) provide (as appropriate) compensation, reparation or restitution to climate displaced persons for damage suffered when that State refrains from protecting and assisting, or fails to protect and assist, climate displaced persons.

## V POST-DISPLACEMENT AND RETURN

*Principle 19: Framework for Return*

(a) States should develop a framework for the process of return in the event that displacement is temporary and return to homes, lands or places of habitual residence is possible and desired by those affected.

(b) States should allow climate displaced persons to voluntarily return to their former homes, lands or places of habitual residence, and should facilitate their effective return in safety and with dignity, in circumstances where such homes, lands or places of habitual residence are habitable and where return does not pose significant risk to life or livelihood.

(c) States should enable climate displaced persons to make a free and informed choice on whether to return to their habitual homes or habitual residence (if possible), to integrate locally or to relocate by:
   (i) consulting such persons on these and other options;
   (ii) ensuring their participation in finding sustainable solutions; and
   (iii) providing such persons with complete, objective, up-to-date and accurate information (including on physical, material and legal safety issues) necessary to exercise their right to decide freely where they want to live.

(d) States should provide transitional assistance to communities during the process of return until livelihoods and access to services are restored.

### *Principle 20: Housing and Livelihood*

(a) States should take appropriate measures, without discrimination of any kind, to allow for the transition by climate displaced persons from temporary or intermediate shelter to permanent housing that fulfills the requirements of "adequacy" in international human rights law, the criteria for "adequacy" being accessibility, affordability, habitability, security of tenure, cultural adequacy, suitability of location, and non-discriminatory access to essential services (for example, health and education).

(b) Where climate displacement results in the ability of climate displaced persons to return to previous sources of livelihood, appropriate measures (for example, re-training) should be taken to ensure such livelihoods can be continued in a sustainable manner and will not result in repeat displacement, and opportunities created by such measures should be available without discrimination of any kind.

## VI AGREEMENT AND IMPLEMENTATION

### *Principle 21: Agreement, Implementation and Dissemination*

In addition to States, communities, non-government organisations and civil society groups at all levels with the ability to agree to, implement and disseminate these Principles should do so without haste.

## Public Consultation Draft
## The Peninsula Principles on Climate Change Displacement

### Table of Contents

## PREAMBLE

ACKNOWLEDGING that events and processes caused or exacerbated by climate change have caused and will continue to cause mass displacement of populations resulting in the erosion of the rights of those affected and the loss of assets, in particular losses relating to housing, land, property and livelihoods;

COGNISANT that virtually all displacement caused by the consequences of climate change is not the responsibility of those suffering such displacement;

NOTING that while climate change displacement can involve both internal and cross border displacement, most climate change displacement will be permanent in nature, occur primarily within State borders, and be caused by a combination of extreme weather events and slow onset environmental change;

CONVINCED, however, that climate change is a global problem, and that the international community has an obligation to assist in resolving that problem through mitigation and adaptation, and to provide assistance to climate change displaced persons regardless of whether the movement of such persons has a cross-border dimension;

REALISING that the international community has obvious humanitarian, social, cultural, financial and security interests in addressing the problem of climate change displacement in a timely, coordinated and targeted fashion;

RECOGNISING that, for many States, addressing the issue of and responding to climate change displacement presents financial, logistical, resourcing and other difficulties, and that there has been no significant coordinated response by States to address human displacement due to climate change, whether domestic or international, temporary or permanent;

RECOGNISING that the United Nations Framework Convention on Climate Change (UNFCCC) and its Kyoto Protocol neither contemplate nor address the issue of climate change displacement, and that conferences and meetings of the parties to these instruments have failed to substantively address climate change displacement;

NOTING, however, that paragraph 14(f) of the COP16 Cancun Adaptation Framework refers to enhanced action on adaptation, including "[m]easures to enhance understanding, coordination and cooperation with regard to climate change induced displacement, migration and planned relocation, where appropriate, at national, regional and international levels";

NOTING, further, that COP18 in Doha decided to establish, at COP19, institutional arrangements to address loss and damage associated with climate change impacts in developing countries that are particularly vulnerable to the adverse effects of climate change as part of the Cancun Adaptation Framework;

RECOGNISING the work being undertaken by numerous United Nations and other intergovernmental agencies to address climate change displacement;

REALISING the need for a globally applicable normative framework to provide principles for the collaborative provision of pre-emptive assistance to those who may be displaced by the effects of climate change, as well as remedial assistance to those who have been so displaced, and legal protections for both;

INSPIRED BY, BUILDING ON AND DRAWING UPON the Nansen Principles on Climate Change and Displacement, the work of the United Nations High Commissioner for Refugees and the UN Principles on Housing and Property Restitution for Refugees and Displaced Persons;

ACKNOWLEDGING the value of normative frameworks such as the Guiding Principles on Internal Displacement (the Guiding Principles), and the use of a number of those Guiding Principles in these Principles, as well as regional initiatives addressing internal displacement such as the African Union Convention for the Protection and Assistance of Internally Displaced Persons in Africa (the Kampala Convention);

REAFFIRMING the strong wish of those who may be displaced by the effects of climate change to remain in their homes and connected to the land on which they live for as long as possible and, where necessary, to move safely and to resettle within their national borders over time, and acknowledging that mechanisms are not currently in place to give effect to that wish;

CONVINCED that climate change-related movement is best understood as a form of adaptation to climate change;

RECOGNISING that government-mandated resettlement has caused impoverishment and social fragmentation and desirous of preventing such consequences as a result of climate change displacement;

RECOGNISING the importance of protecting the collective and individual rights of climate change displaced persons and the need to foster the principle of self-determination;

RECOGNISING the writings of eminent jurists as a source of international law, and acknowledging the importance of such writings in drafting these Guiding Principles on Climate Change Displacement; and

ACKNOWLEDGING comments made on earlier drafts of these Guiding Principles on Climate Change Displacement by scholars and legal practitioners from many jurisdictions, these Guiding Principles on Climate Change Displacement ('Principles') provide as follows:

## I INTRODUCTION

### *Principle 1: Framework*

(a) These Principles provide a comprehensive, normative framework, based on principles of international law and best practice, within which the rights and needs of climate change displaced persons can be addressed. They set out protection and assistance principles to be applied to such persons.
(b) After setting out principles of general application, these Principles address:
   (i) climate change displacement prevention and planning;
   (ii) provision of climate change displacement assistance and protection, including remedies and compensation;
   (iii) post-displacement return and resettlement; and
   (iv) climate change displacement across national borders.

### *Principle 2: Definition of Climate Change Displaced Persons*

1. "Climate change displaced persons" in these Principles mean persons or groups of persons whose homes or places of habitual residence have or may become temporarily or permanently uninhabitable as a result of the effects of climate change and whose housing, land and property rights may be under threat.
2. As defined in Principle 2(1), climate change displaced persons include people who, in order to protect their safety, rights and dignity, may need or wish to be resettled:
   (a) from areas prone to extreme weather events, slow-onset environmental change and the combination of both which threaten lives and livelihoods; or
   (b) because their lands are needed for mitigation and/or adaptation measures.

### *Principle 3: Implementation of Domestic Procedures, Institutions and Mechanisms*

(a) States should establish and support equitable, timely, independent and transparent procedures, institutions and mechanisms – at the tribal and local/community group level to the national level – to implement these Principles and give effect to their provisions, as more fully set out at Principle 12.
(b) In particular, States should incorporate climate change displacement assistance and protection provisions as set out in these Principles, including voluntary migration and resettlement due to the effects of climate change, into domestic law and policies, and provide adequate resources for their implementation.
(c) Where existing procedures, institutions and mechanisms can effectively implement these Principles and give effect to their provisions, adequate resources should be made available to facilitate that implementation.

### *Principle 4: No Discrimination against Climate Change Displaced Persons*

States should not discriminate among climate change displaced persons, and climate change displaced persons shall not be discriminated against on the basis of their displacement or race, colour, gender, sexual orientation, language, religion or belief, political or other opinion, national, ethnic or social origin, legal or social status, age, disability, property, birth, or on any other similar grounds.

### *Principle 5: Interpretation*

These Principles shall not be interpreted as limiting, altering or otherwise prejudicing the rights recognized under international human rights, refugee and humanitarian law and related standards, or rights consistent with those laws and standards as recognized under domestic law.

## II GENERAL PRINCIPLES

### *Principle 6: State-Based Provision of Adaptation Assistance and Protection*

States should, individually or collectively, provide adaptation assistance and protection to climate change displaced persons so that such persons can remain in their homes for as long as possible or can move within their state or across borders in a planned manner over time.

### *Principle 7: State Cooperation in Provision of Adaptation Assistance and Protection*

Climate change induced displacement is a matter of global responsibility, and States should be obliged to cooperate in the provision of adaptation assistance and protection to climate change displaced persons whether or not such displacement occurs within a State or across national borders.

## III CLIMATE CHANGE DISPLACEMENT PREVENTION AND PLANNING

### *Principle 8: Risk Management, Monitoring and Modeling*

States should:

(a) identify options and design and implement risk management strategies and approaches, including risk reduction and risk-transfer and risk-sharing mechanisms, in relation to climate change displacement;

(b) undertake systematic observation and monitoring of, and data collection at the local, regional and national levels on, climate change displacement;
(c) enhance sharing, access to and the use of such data at the local, regional and national levels to facilitate the assessment and management of climate change-related displacement; and
(d) model likely climate change displacement scenarios, locations threatened by climate change, and possible resettlement sites for climate change displaced persons.

### *Principle 9: Community Preparation and Planning*

States should:

(a) integrate resettlement due to climate change within national laws and policies;
(b) develop an institutional framework to:
   (i) identify indicators that will cause planned resettlement and allow for government technical assistance and funding; and
   (ii) outline steps a community must take prior to climate change displacement in order to receive such technical assistance;
(c) to the extent necessary, through consultation, education and awareness raising, prepare at-risk and potential host communities for the prospect of climate change displacement, and what such displacement involves;
(d) in planning for displacement due to climate change, include measures to promote livelihoods and economic prosperity and minimize risks in relation to both displaced and host communities;
(e) ensure that, in terms of preparing for climate change displacement:
   (i) affected communities (both displaced and host) are informed, consulted and can participate in relevant decisions and the implementation of those decisions, including the planning and implementation of housing and resettlement programmes;
   (ii) basic services, adequate and affordable housing and access to livelihoods (without discrimination) are available for climate change displaced persons in the receiving community;
   (iii) adequate mechanisms and safeguards are in place to prevent and solve conflicts over land and resources; and
   (iv) the rights of climate change displaced persons are protected before, during and after the displacement; and
(f) prior to any resettlement, prepare a master resettlement plan which takes account of matters including:
   (i) land acquisition;
   (ii) community preferences;
   (iii) transitional shelter and permanent housing;

(iv) the preservation of existing social and cultural institutions of climate change displaced persons;
(v) access to public services;
(vi) support needed during the transitional period;
(vii) family and community cohesion;
(viii) concerns of the host community;
(ix) monitoring mechanisms; and
(x) grievance procedures.

*Principle 10: Land Identification, Habitability, Use and Resettlement*

(a) States should, in particular, afford protection against climate change displacement to those persons and groups in their territory who are particularly dependent on and attached to their land, including indigenous people and those reliant on customary rules relating to the use and allocation of land.
(b) States should, in the course of planning for and anticipating climate change displacement and the voluntary resettlement of communities:
    (i) set aside sufficient public and other land as may be required to provide land-based solutions to all climate change displaced persons;
    (ii) develop a fair and just land acquisition process and appropriate land resettlement and allocation programmes;
    (iii) plan for and develop new settlements and towns on land not at risk from the effects of climate change and, in so planning:
        (A) take measures to ensure that the land chosen for resettlement is not vulnerable to future climate change-related hazards; and
        (B) consider the safety and environmental integrity of the new site(s) and ensure that the rights of both those resettled and the communities which host them are upheld;
    (iv) develop proactive policies and programs (including appropriate incentives, assistance and funding) to encourage those likely to be displaced as a result of the effects of climate change to move before they are affected; and
    (v) take measures such as land audits, demarcation of uncontested boundaries, and community land mapping to facilitate the identification of land for when people need to be temporarily or permanently moved, either internally or across borders, as a result of the effects of climate change.
(c) More particularly, in order to determine the habitability and feasibility of any resettlement site, and to ensure that climate change displaced persons being resettled and the relevant State are in agreement as to the habitability of the relevant resettlement site, States should create specific, geographically relevant, standard criteria including:
    (i) current land use;

(ii) restrictions (including those of a customary nature or not otherwise formally codified) associated with the land and its use; and
(iii) habitability of the land, including issues such as accessibility, availability of water, vulnerability to climate hazards, and feasibility of subsistence/agricultural use, together with mechanisms for communities to decide to where they wish to resettle.

(d) States agree that climate change displaced persons have the right to participate fully in decisions with regard to their voluntary resettlement (for example, whether they resettle, destination, timing, the manner in which the movement is to be effected, housing options, livelihood and economic issues) and to be provided with easily accessible information concerning:
(i) the nature and extent of the changes to land and habitats resulting from climate change, including the evidence on which such assessments are made;
(ii) evidence that all other alternatives to resettlement have been considered (including mitigation and adaptation measures that could be taken to enable people to remain in their communities);
(iii) planned efforts to assist affected communities in resettling, both short and long-term;
(iv) available compensation and alternative resettlement options if a choice is made not to be resettled under the Government's planning; and
(v) rights of climate change displaced persons under international and domestic law.

(e) States should include in resettlement planning:
(i) measures to compensate climate change displaced persons for lost housing, land and property; and
(ii) assurances that the housing and livelihoods needs are met of those climate change displaced persons who have informal land rights, customary land rights, occupancy rights or rights of customary usage.

(f) States should consider the necessity of amending national laws to ensure that the rights of minority or other vulnerable groups (including women, the elderly and indigenous populations) are respected.

### *Principle 11: Loss and Damage*

(a) States should develop appropriate laws and policies on compensation for loss suffered and damage incurred by climate change displaced persons.

(b) States should, more particularly:
(i) assess the risk of loss and damage associated with climate change displacement;
(ii) identify and develop appropriate approaches to address loss and damage associated with climate change displacement, including through risk reduction;

(iii) undertake systematic observation and monitoring of, and data collection at the local, regional and national levels on, the impacts of climate change, and account for likely and actual losses, as appropriate, in terms of climate change displacement;
(iv) enhance understanding of how loss and damage associated with the adverse effects of climate change affects those segments of the population already vulnerable to displacement;
(v) examine how approaches to address loss and damage associated with climate change displacement may be integrated into climate-resilient development processes; and
(vi) involve communities and populations vulnerable to displacement, civil society, the private sector and other relevant stakeholders in the assessment of and response to loss and damage.

*Principle 12: Institutional Framework to Support and Facilitate the Provision of Assistance and Protection*

(a) States should strengthen national capacities and capabilities to identify and address the protection and assistance needs of those vulnerable to or already suffering from climate change displacement.
(b) States should take all appropriate administrative, legislative and judicial measures, including the creation of adequately funded Ministries, departments, offices and/or agencies at the national, regional and local levels empowered to:
  (i) develop, establish and implement an institutional framework to:
    (A) identify the indicators that will cause climate change displacement;
    (B) enable government technical assistance and funding and to outline the steps a community must take prior to displacement in order to receive such technical assistance; and
    (C) support and facilitate the provision of assistance and protection to climate change displaced persons; and
  (ii) ensure full implementation of these Principles.
(c) Responsibility for establishing such departments, offices and/or agencies should lie with national governments, and such governments should consult and collaborate with state, regional and local governments (or authorities), as appropriate, in this regard, and integrate such departments and agencies in such institutional framework.
(d) States should ensure the provision of adequate resources (including points of contact and assistance) at the level of government which would most directly address the concerns of climate change displaced persons.
(e) The institutional framework referred to in this Principle 12 should address matters set out in this Part III and Part VI of these Principles.

### *Principle 13: Bilateral and Regional Arrangements between States*

States should consider entering into bilateral and regional arrangements with other States as a coordinated means of preventing and planning for climate change displacement.

### *Principle 14: International Law Obligations*

States should, in all circumstances, ensure respect for their obligations under international law so as to prevent and avoid conditions that might lead to displacement of persons due to the effects of climate change.

## IV CLIMATE CHANGE DISPLACEMENT

### *Principle 15: State Cooperation in Assisting and Protecting Climate Change Displaced Persons*

(a) States should cooperate with each other in assisting and protecting climate change displaced persons.
(b) Affected States should consider in good faith offers of assistance and support from other states and from humanitarian and like organisations, whether made individually or collectively, in relation to climate change displacement.
(c) States should allow quick and unimpeded entry and passage of humanitarian personnel and aid and assistance to climate change displaced persons.
(d) States should enable and facilitate the role of tribal, local, State and international organisations and humanitarian agencies, civil society organisations and other relevant actors in providing assistance to climate change displaced persons.
(e) States should create and maintain databases of climate change displaced persons within either their jurisdiction or effective control and, in so doing, collaborate with international organisations and agencies and civil society organisations.

### *Principle 16: Inability of States to Provide Assistance*

A State should, if it is unable (for whatever reason) to provide assistance to climate change displaced persons no matter whether such persons are displaced within that State or across borders, seek assistance and cooperation from other States and relevant international agencies, and such other States and agencies, either separately or together, should provide such assistance and cooperation to that State.

### *Principle 17: State Assistance to Climate Change Displaced Persons*

States should provide climate changed displaced persons under their jurisdiction (whether displaced internally or across borders) with a practicable level of age and gender-sensitive humanitarian assistance including, without limitation:

(a) emergency services;
(b) evacuation and resettlement;
(c) medical assistance and other health services;
(d) shelter;
(e) food;
(f) clean water;
(g) sanitation;
(g) measures necessary for social and economic inclusion including, without limitation, anti-poverty measures, free and compulsory education, training and skills development; and
(h) facilitation of family reunion.

*Principle 18: Rights, Freedoms and Non-Discrimination*

Climate change displaced persons shall:

(a) enjoy, in full equality, the same rights and freedoms under international and domestic law as do other persons in their country; and
(b) not be discriminated against in the enjoyment of any rights and freedoms on the ground that they are internally displaced.

*Principle 19: Legal, Economic, Social and Other Protection Measures*

(a) States have the primary duty and responsibility to provide all necessary legal, economic, social and other forms of protection and assistance to those climate change displaced persons displaced within their borders (and to those likely to be displaced within their borders) due to climate change.
(b) States are responsible for ensuring that their actions in terms of protection and assistance are consistent with both domestic and international law.
(c) Protection and economic activities undertaken by States should be carried out in a manner that respects both the cultural Sensitivities prevailing in the affected area and the principles of maintaining family and community cohesion.
(d) As climate change displacement is a matter of global responsibility, the international community should cooperate in the provision of protection and assistance to those displaced within a State by climate change in the event that such State is unable to provide protection and assistance.

*Principle 20: Prohibition of Arbitrary and Discriminatory Laws*

(a) States should neither adopt nor apply laws that prejudice the rights, freedoms and needs of climate change displaced persons being addressed.

(b) States should take steps to repeal unjust or arbitrary laws and laws that otherwise discriminate against, or have a discriminatory effect on, climate change displaced persons.

*Principle 21: Remedies and Compensation*

States should:

(a) ensure that climate change displaced persons are entitled to and should be supported in claiming and exercising their rights and provided with effective remedies, including unimpeded access to the justice system, including in case of violations or when conflicts emerge within or between communities affected by the resettlement;
(b) establish legal frameworks to provide compensation and reparation (where appropriate) on fair and just terms to climate change displaced persons; and
(c) make reparations to climate change displaced persons for damage suffered when that State refrains from protecting and assisting, or fails to protect and assist, climate change displaced persons.

## V POST-DISPLACEMENT RESETTLEMENT AND RETURN

*Principle 22: Framework for Resettlement and Return*

States should develop a framework for the processes of (a) resettlement; and (b) return, in the event that resettlement is temporary and return to homes, lands or places of habitual residence is reasonably possible; and the operation of those processes, with regard to climate change displaced persons, such framework to incorporate Principles 9 and 10 and Principles 23 to 26.

*Principle 23: Domestic Return, Local Integration or Resettlement*

(a) States should allow climate change displaced persons to voluntarily return to their former homes, lands or places of habitual residence (whether such persons have been displaced internally or across borders), and should facilitate their effective return in safety and with dignity, in circumstances where such homes, lands or places of habitual residence are habitable and where return does not pose significant risk to life or livelihood.
(b) States should enable climate change persons displaced internally to make a free and informed choice on whether to return to their habitual homes or habitual residence (if possible), to integrate locally or to resettle by:
  (i) consulting such persons on these and other options;

(ii) ensuring their participation in finding sustainable solutions; and
(iii) providing such persons with complete, objective, up-to-date and accurate information (including on physical, material and legal safety issues) necessary to exercise their right to decide freely where they want to live.

(c) States should take appropriate measures to provide the means to enable climate change displaced persons to:
(i) return to their habitual homes;
(ii) integrate locally; or
(iii) resettle.

(c) States should provide transitional assistance to communities during the process of resettlement until livelihoods and access to services are restored.

*Principle 24: Housing and Livelihood*

(a) States should take appropriate measures, without discrimination of any kind, to allow for the transition by climate change displaced persons from temporary or intermediate shelter to permanent housing, fulfilling the requirements of "adequacy" in international human rights law, the criteria for "adequacy" being accessibility, affordability, habitability, security of tenure, cultural adequacy, suitability of location, and access to essential services (for example, health and education).

(b) Where climate change displacement results in the inability of climate change displaced persons to return to previous sources of livelihood, appropriate measures (re-training, for example) should be taken, and opportunities created by such measures should be available without discrimination of any kind.

*Principle 25: Housing and Property Restitution, and Compensation*

(a) States should prioritise the right to restitution as the primary remedy for climate change displacement.

(b) States should
(i) restore to climate change displaced persons housing, land and/or property of which they were deprived; and/or
(ii) compensate climate change displaced persons for housing, land and/or property that is, as a matter of fact, impossible to restore.

(c) States should adopt positive measures aimed at alleviating the plight of climate change displaced persons living in inadequate housing.

(d) States should take all appropriate administrative, legislative and judicial measures to support and facilitate housing, land and property restitution processes and procedures.

## VI PRINCIPLES ON CROSS-BORDER CLIMATE CHANGE DISPLACEMENT

### *Principle 26: Cross-Border Assistance*

(a) States (individually and acting together) should consider what assistance can be provided to those either displaced or who may be displaced (including legal recognition and protection for those displaced or likely to be displaced) across borders by climate change).
(b) In particular, States should consider strengthening education and training for those either displaced or who may be displaced such that they could migrate with dignity if they so chose.

### *Principle 27: Respect for Human Rights and Minimum Standard of Treatment*

States should ensure full respect for, and protect and ensure, the human rights of those displaced across borders as a result of the effects of climate change.

### *Principle 28: Eligibility Criteria for Assistance and Protection*

Climate change displaced persons displaced across national borders should remain eligible for assistance and protection until they acquire a new nationality, are voluntarily returned to their home country, or refuse to return when it is safe for them to do so.

### *Principle 29: Non-Refoulement*

States should not forcibly return any climate change displaced person to his or her home state in situations where such person's life or freedom could be threatened.

### *Principle 30: Climate Change Displacement Agreements between States*

(a) Principles of proximity, self-determination and the safeguarding of intangible culture should guide both decision-making by "home" and "host" states in the event that climate change displacement agreements are made between such states for the resettlement or movement of climate change displaced persons.
(b) States should review existing migration and immigration policies to allow those displaced – or likely to be displaced – by climate change to voluntarily migrate to another State.

# Index